Routledge Revivals

Paradoxes of the Infinite

Paradoxes of the Infinite presents one of the most insightful, yet strangely unacknowledged, mathematical treatises of the 19th century: Dr Bernard Bolzano's *Paradoxien*. This volume contains an adept translation of the work itself by Donald A. Steele S.J., and in addition an historical introduction to the masterpiece, which includes a brief biography as well as an evaluation of Bolzano the mathematician, logician and physicist.

Paradoxes of the Infinite

Bernard Bolzano

Routledge
Taylor & Francis Group

First published in 1950
by Routledge & Kegan Paul Ltd

This edition first published in 2014 by Routledge
2 Park Square, Milton Park, Abingdon, Oxon, OX14 4RN
and by Routledge
711 Third Avenue, New York, NY 10017
Routledge is an imprint of the Taylor & Francis Group, an informa business

A Library of Congress record exists under LC control number: 51001079

ISBN 13: 978-0-415-74976-3 (hbk)
ISBN 13: 978-1-315-79578-2 (ebk)

PARADOXES OF THE INFINITE

by Dr. Bernard Bolzano

TRANSLATED FROM THE GERMAN
OF THE POSTHUMOUS EDITION
BY DR. FR. PŘIHONSKÝ
AND FURNISHED
WITH A HISTORICAL INTRODUCTION
BY DONALD A. STEELE
S.J., D.Ph.

LONDON

ROUTLEDGE AND KEGAN PAUL

This edition first published in 1950
by Routledge and Kegan Paul Ltd.
68 Carter Lane, London E.C.4
and printed in Great Britain
by William Clowes & Sons Ltd.
Beccles
Typography by Seán Jennett

De Licentia Superiorum Ordinis

Tender-hearted stroke a nettle,
And it stings you for your pains;
Grasp it like a man of mettle,
And it soft as silk remains.

If there be in mathematics a nettle danger out of which has been plucked the flower safety, it is speculation on O and ∞.

Augustus De Morgan, *Transactions of the Cambridge Philosophical Society*, 11 (signed 28 February 1866), 258. The quatrain is taken by De Morgan from the *Verses written on a Window in Scotland*, by Aaron Hill (1685–1750).

CONTENTS

HISTORICAL INTRODUCTION

(1) *means: Footnote* 1 *at end of current division.* (B1) *means: Entry* 1 *in the Bibliography, pages* 176–81. (B1, 1) *means: page* 1 *of entry* 1.

TRANSLATION

Asterisks mark footnotes translated currently from Bolzano. (1) *means: Translator's note* 1 *at the end of the Translation.*

SELECTED BIBLIOGRAPHY

SHORT TITLE KEY TO BOLZANO REFERENCES

BD=*Beiträge zu einer begründeteren Darstellung*, Prague, 1810, 152 pp.

BG=*Betrachtungen über einige Gegenstände*, Prague, 1804, 63 pp.

BL=*Binomischer Lehrsatz*, Prague, 1816, 144 pp.

BS=*Begriff des Schönen*, KBGW (5) 3, 1845, 1–92.

DD=*Drei Dimensionen*, Prague, 1843, and KBGW, 15 pp.

DP=*Drei Probleme*, Leipzig, 1817, 80 pp.

EK=*Einteilung der Künste*, KBGW (5) 6, 1851, 133–178.

KBGW=*Abhandlungen der königlich-böhmischen Gesellschaft der Wissenschaften*, Prague.

PU=*Paradoxien des Unendlichen*, Leipzig, 1851, 134 pp.

RB=*Rein analytischer Beweis*, Prague, 1817, and KBGW, 60 pp.

WL=*Wissenschaftslehre*, Sulzbach, 1837, four volumes, 2397 pp.

ZK=*Zusammensetzung der Kräfte*, Prague, 1842, and KBGW, 40 pp.

NOTE ON REFERENCES TO THE 'WISSENSCHAFTSLEHRE'

The Paragraphs are numbered through four volumes as follows: §§1–120 in I, §§121–268 in II, §§269–391 in III, §§392–718 in IV.

For fuller descriptions see the Bibliography on pages 176–9.

HISTORICAL INTRODUCTION

I. The Historical Context of the *Paradoxien*

BLENDING precision with temperament and a keen scent for fresh sources, Heinrich Scholz is not content in his *Geschichte der Logik* (B43) with the annals of terminology and technique. The first of his three parts is devoted to the changing 'shape' of logic as a whole, to its changing degree of formalism, to its changing frontiers with psychology and epistemology. The historian of mathematics has learnt to do likewise. He has learnt to shift the stress from anecdotes, priorities and descriptions to the emergence, the evolution and the interaction of concepts, problems, methods and theory structures. Two of the tasks to be undertaken in this spirit are, to trace the changing standard of rigour in mathematics and to trace its changing frontiers with physical science.

Hazardous though division into periods is, and doubly hazardous one by centuries, we reach a passable first approximation if we regard the centuries of Eudoxus and Archimedes, of Weierstrass and Hilbert as four centuries of greater rigour, and the century of the Enlightenment and the Encyclopedia as one century of greater laxity, and indeed of a laxity beyond the simple lack of conscientiousness in definition and demonstration. Figures of transition will offer the historian an interest of their own—not only figures like Gauss, the lonely giant, austere and integral, whose Minerva-born precision in real as well as complex function theory only became known long after his death, but also those other figures in whose minds the battle for logical ideals swayed to and fro, with losses as well as gains. Such a figure was De Morgan (1806–1871) in London and such another was Bernard Bolzano (1781–1848) in Prague.

Through Leibniz and Newton and other substantial contributors, and not least through the resuscitated influence of Eudoxus and Archimedes, powerful new methods brought mathematical analysis into being: yet these methods could not be pursued at once to their logical foundations or to their full generality. They were applied instead to particular problems inspired by the growth of physical science. Intoxicating success in this field not only tended to relax logic, but also to mask problems of existence and uniqueness. It was hard at that time, taking a simple example, to see in velocity not a physical entity reached by our senses, but an indirect and sophisticated construction, legitimate and significant to be sure, but only at the price of the same disciplined reflection which fashioned at last, with diamond-cutter's patience, the four Dini derivates of a function. It was harder to understand, in consequence, that 'velocities' ingenuously accepted as existent and bilateral are no secure basis for a 'differential calculus' (1).

Such seeming generality as lay in the use of infinite series tempted even Newton himself to speak of 'aliarum omnium quadratura' (2), and was a stronger temptation still to Euler, whose adventurous computations, checked by transient scruples, produced numerous results that happened to be true and very valuable and eventually demonstrable, together with other results patently absurd. At length, the genius who inverted elliptic integrals also saw that the bare correction of fallacies was necessary but not sufficient. One '*Fandenskab*,' one impish puzzle, remained to stimulate Abel: why had the invalid methods not more often led to false results (3)? Precisely the same question crossed the mind of Bolzano (4) and led him to do his own work on some of the fundamental problems of mathematics.

Nor was laxity in ratiocination the only evil inviting reform at the hands of Bolzano and Abel, of Gauss more privately and deeply, of Cauchy and Weierstrass more publicly and efficaciously. The mere hardihoods of Euler (5) are a commonplace of mathematical history; but he went farther. After defining a function for integer arguments, he tried to 'deduce' its value for fractional ones by a quasi-magical interpolation (6). He openly advises the reader to trust his calculations rather than his judgement (7) and makes in the *Mechanica* a suggestion best regarded as a joke at his own expense: on arriving at a point where mathematical difficulties occur, a particle is blandly said (8) to suffer annihilation! Mathematics might, in short, be a world of natural phenomena parallel to that of physics, with entities awaiting discovery as America awaited Leif Ericsson or Columbus. The identification of $\log(-1)$ was hotly disputed before ever the function $\log(x)$ had been defined over $x=-1$ at all, and this sterile controversy (9) was only laid to rest by Gauss (10), though then very thoroughly. Even the sense of aptness in notation must have been dull when D'Alembert denoted (11) the first three derivates of $f(x)$ by $\Delta(x)$, $\Gamma(x)$ and $\Psi(x)$. The genial poet-mathematician or mathematician-poet Abraham Gotthelf Kaestner wrote more truly than he realised when, with a touch of Hogarth and a sly quotation from Ovid, he compared the analysis of his day (12) to a beautiful maiden clad in slatternly garments; and that nursing-mother of subtleties in algebra and analysis, and of ideas so beautiful and fertile and ordinative as those of group and ring and ideal, the theory of numbers, suffered the gibe of Voltaire (13) harmlessly but not unrepresentatively.

Nevertheless, history is not discontinuous enough for the eighteenth century to lack misgivings. Within three years Benjamin Robins (14) castigated the vagaries quoted from Euler's *Mechanica* and a third vagary which our Bolzano was also to reject (PU, §70), if for reasons of his own. The terms 'convergent' and 'divergent' had been introduced by James Gregory (15), one definition had been interestingly refined by Robert Simson (16), Varignon had sobered Leibniz' quasi-magical summation of the Grandi series (17) and Leibniz himself, questioning Newton's right to use series so freely, openly desiderated criteria of 'advergence' (18), now called 'convergence.' Euler himself addresses reproaches to those who neglect this or that precaution (19). Laplace begins (20) to protest at the summation of the Grandi series by 'probability considerations,' Clairaut begins to rate faults of procedure as more serious than faults of calculation and Lagrange begins (21) to demur at uncontrolled approximations and the failure to estimate remainders. For one clouded moment in which D'Alembert says 'allez en avant, la foi vous viendra' (22), we have two lucid ones in which the same D'Alembert says that simplicity and rigour are not mutually exclusive but mutually helpful, and that results obtained from divergent series remain suspect even when they harmonise with facts otherwise known (23).

Quietly, in the last year of that century, three presage-laden sheets were being written, but were to remain unprinted until 1912. Starting with a bounded sequence (a_n) the writer arrived by clearcut steps and one then pardonable gap at a number L_0 such that each higher number was exceeded by no term, and each lower number by at least one term. Successive termwise decapitations then gave monotone sequences of 'fines superiores' (L_n) and corresponding 'inferiores' (M_n), whereupon the former had in turn a 'finis inferior' L and the latter a 'superior' M. When L and M coincided, the original sequence was convergent and had their common value for its hitherto cruder 'limit.' Notwithstanding the gap and the absence of the modern Latin names, we here stand closer to the 'limites superior et inferior' than we do in the *Cours d'Analyse* of Cauchy (24). These three sheets (the turn of the tide back to logic and so back from manipulation to mathematics) were the work of Gauss (25) at the age of twenty-four; but to notice the selfsame gap, and to attempt its bridging, was the work of Bolzano.

When Gauss in 1812 deliberately caters for lovers of rigour and appeals with justice to Greek ideals, initiates deeper convergence testing, warns against the snags of divergence and plans a dissertation expressly on that theme (26); when Abel in 1826 corrects Cauchy's belief that a sequence of continuous functions must needs have, if any, then a continuous limit function, dispenses with absolute convergence at a boundary and uses uniform convergence even if without isolating it as a concept (27); when Dirichlet warns Steiner that sundry 'identifications' of maxima repose on existence hypotheses (28), then the great watershed is fairly behind us. And more recent, more competent examinations of Bolzano's mathematical remains reveal that he, like Gauss, was eremitically and uninfluentially yet distinctly ahead of his contemporaries in surprisingly many of these delicate points.

The battle for logical conscientiousness (not as barren

pedantry, but for rigour in mathematics as the shield of its autonomy, the spring of its fruitfulness, the condition of its beauty) was now joined. Weierstrass and Cantor, Kronecker and Frege, Brouwer and Weyl could carry it on each in his own time and fashion. Whitehead could look back (29) on the easier satisfaction of our mathematical forefathers, Russell could dismiss infinitesimals (30) with four scarifying adjectives and Hardy's *Pure Mathematics* could preach the epsilontic gospel to Littlewood's 'cannibals.' But victories came slowly at first. The logically and hellenistically preciser research on Greek mathematics published during the short career of *Quellen und Studien* (31) will best prepare the reader to savour the exquisite shallowness of Lacroix's dictum (32) that 'we have no further need of the subtleties with which the Greeks plagued themselves'! Subtleties which touch again and again on problems but recently rediscovered! After learning from Bolzano in 1817, save for a real number preconstruction, that $|a_m - a_n| < \varepsilon$ for all $m > n > \nu(\varepsilon)$ is sufficient as well as necessary for the convergence of (a_n), we are somewhat disconcerted to find Laurent in 1862 bereft of counsel (33) at this point: not missing what we miss, but muttering that we must bypass an 'obstacle which it would be too difficult to overturn.' Again, when Bertrand's *Traité* of 1864 simply takes the convergence of a bounded monotone sequence for granted (34), our minds travel involuntarily forty-seven years back to the treatise in which a proof was attempted with insight and courage, and achieved less incompletely than was to be expected: namely, to the *Rein analytischer Beweis* of Bolzano. Two years afterwards, in 1866, the same climate reigns in Duhamel as in Bertrand, and a work professing by title to discuss the 'methods of the sciences of ratiocination' proves the Bolzano-Weierstrass Theorem for large intervals by the simple device of assuming it for smaller ones (35). Even De Morgan, amid many sallies towards rigour, is once so exuberant in language as to say that he could believe anything of a function which became infinite (36): presumably less a counsel of despair than a humorous warning to expect the unexpected.

In brief, it would be a simplification unjust in one direction and unmerited in the other if we qualified the eighteenth century, for all its fecundity in verificanda, as one of unrelieved laxity in mathematical reasoning, and the nineteenth as one of unsullied rigour. Nevertheless, the transition was markedly in that sense. Euler and D'Alembert, Boscovich and Hindenburg, Kaestner and Laplace belong to the one epoch, and Legendre with Lagrange prepare the next: and all these mathematicians died during Bolzano's lifetime. Cauchy and Jacobi, Dirichlet and Weierstrass, Hamilton and De Morgan, Grassmann and Boole, Riemann, Kronecker and Dedekind, Dini and Cantor belong to another and very different epoch: and all these mathematicians were born during Bolzano's lifetime. When we reflect too that Daniel Bernoulli's life just overlapped into Bolzano's, that the life of Gauss embraced it, that Bolzano's life embraced those of Abel and Galois and just overlapped into that of Frege, we see that even in the bare chronological sense Bolzano marks a distinct point of historical inflection. That he does so with his actual work in mathematics as well must be shown after we have given a short sketch of his life.

NOTES TO I: THE HISTORICAL CONTEXT OF THE *PARADOXIEN*

(1) An opposite and no less characteristic view underlies the anecdote of the student reproved for defining a derivate as a limit and not as a velocity. See Alexander Russell: *Lord Kelvin.* London and Glasgow 1938, pp. 30–31.

(2) *Analysis per aequationes numero terminorum infinitas.* Extant 1669. published 1704. See heading of Chapter III.

(3) Letter to Holmboe, 16 January 1826.

(4) *Paradoxien des Unendlichen*, §37, p. 65.

(5) In the *Institutiones Calculi Differentialis* (1775), 500, the summatory formula for a series differs according as n is odd or even, but despite the reference to different n, successive couples are added, n apparently eliminates itself and the average is declared to be the sum to infinity. Another drastic example occurs in §19 of *Considerationes de seriebus quibusdam*, Werke (1) 14 (1925), 427.

(6) *Sur le temps de la chute d'un corps attiré vers un centre de forces en raison réciproque des distances.* Histoire de l'Académie de Berlin (1760), 250–260. See p. 256.

(7) *Mechanica I*, 108, Werke (2) 1 (1912), 88.

(8) *Mechanica I*, 315, Werke (2) 1 (1912), 258–259.

(9) The literature of this notorious controversy on the logarithms of negative and complex numbers fills an entire page in F. W. A. Murhard's *Bibliotheca Mathematica*, 2 (1798), 286; and a particular account of it was composed by B. F. Thibaut, Göttingen, 1797.

(10) Letter to Bessel, 18 December 1811, Auwers, p. 157. Gauss defines the logarithmic function as a complex integral and explains its multiple value by differing paths of integration.

(11) *Recherches sur différents points importants du systême du monde.* Paris 1 (1754), 50.

(12) *Anfangsgründe der Analysis des Unendlichen.* Göttingen, 1799. See the peroration to the preface, signed March 1761.

(13) *Dictionnaire philosophique*, article 'Géométrie,' at end.

(14) *Remarks on Mr. Euler's Treatise of Motion.* London 1739, pp. 1–29 on Euler, pp. 30–112 on others named in title.

(15) *Vera circuli et hyperbolae quadratura*, Padua 1668.

(16) *De limitibus quantitatum et rationum fragmentum.* Opera quaedam relicta, Glasgow 1776. See pp. 3–4.

(17) Letter to Leibniz, 19 November 1712. Gerhardt, *Leibnizens mathematische Schriften*, 4 (1859), 188.

(18) Letter to Johann Bernoulli, 25 October 1713. Gerhardt, *Leibnizens mathematische Schriften*, 3 (1885), 922–923.

(19) Opera posthuma, edited by F. H. and N. Fuss. Petrograd 1 (1862), 301.

(20) *Théorie des probabilités.* Paris 1820, p. cix.

(21) *Théorie des fonctions.* Paris 1796, §53.

(22) J. Bertrand: *D'Alembert.* Paris 1889, p. 56.

(23) *Opuscles mathématiques.* Paris 5 (1768), 183. The notion that Cesàro and Hölder and similar means be methods of capturing an elusive but pre-existent limit is a relic of eighteenth-century misplaced concreteness.

(24) Paris 1821, Chapter VI, pp. 114–152. Oeuvres (2) 3 (1897). The entry of 'la plus grande des limites' is, in comparison with Gauss, rather surreptitious.

(25) L. Schlesinger: *Ueber Gauss' Arbeiten zur Funktionentheorie.* Materialen für eine wissenschaftliche Biographie von Gauss, Heft III. Berlin, 1912. Full text on pp. 136–140.

(26) *Disquisitiones generales circa seriem hypergeometricam.* Göttinger Kommentarien 1812, Werke 2 (1876), 125–162. See in particular pp. 139 and 152.

(27) In his memoir on the binomial series, Crelle 1 (1826), 311–339. See pp. 316 and 337.

(28) E. Lampe: *Zur Biographie von Jakob Steiner.* Bibliotheca Mathematica, (3) 1 (1899), 129–141. See p. 134.

(29) *Introduction to Mathematics*. London 1919, pp. 155–157.

(30) 'Irrelevant, unnecessary, erroneous and self-contradictory': *Principles of Mathematics*. Cambridge 1903 (reprinted 1938), p. 345.

(31) *Quellen und Studien zur Geschichte der Mathematik, Astronomie und Physik*. Berlin (Springer) 1–4 (1929–1938). Contains some twenty-five historiographically rigorous studies in Greek mathematics by O. Becker, J. Klein, O. Neugebauer, J. Stenzel, O. Toeplitz and others.

(32) *Traité de calcul différentiel*. Paris 1 (1810), 11.

(33) *Théorie des séries*. Paris 1862, p. 6.

(34) *Traité de calcul différentiel*. Paris 1864, §229.

(35) J. M. C. Duhamel: *Des méthodes dans les sciences du raisonnement*. Paris 2 (1866), 413.

(36) *Transactions of the Cambridge Philosophical Society*, 11 (signed 28 February 1866), 242.

II. Biographical Sketch

BERNARD PLACIDUS JOHANN NEPOMUK BOLZANO was born on 5 October 1781 in the Old Town of Prague, as the fourth child of Bernard Bolzano, an art dealer who had migrated thither from Nesso in Italy. To his mother, born Cecilia Maurer, he paid ardent but carefully reasoned tribute in the Autobiography (B6); she had reared him to frugal self-reliance and sound intellectual habits, shared his academic mischances and patiently borne the loss of husband and ten children. As a boy, Bolzano was weak and ailing, nervous and excitable, subject to unremitting headaches and thus cut off from his natural playmates. The Autobiography owns to false shyness and reserve, to fits of sudden anger suddenly stilled and to juvenile contentiousness. In riper years Bolzano took anxious and rather scrupulous pains to correct the faults he discerned in himself, but his bodily health, his eyesight and his musical ear remained poor to the end of his life.

During his five years at the classical academy of the Piarists Bolzano was neither a brilliant nor a highly placed pupil, and on entering the University of Prague in 1796 he found philosophy 'almost as difficult as mathematics.' Indeed, the future author of the *Wissenschaftslehre* found its hardest branch to be—logic. Books were scarce and dear, whereupon Bolzano comments wryly and drily that he was thus forced to think the more. That the gain in power of speculation was offset by weakness in the exact humanities and in positive theology is a fact which he came to realise, to regret and in part to repair by wide reading.

The academic mathematics of the period at first repelled Bolzano, whose spirit was more strongly kindled by reading the greater and lesser masters. At an early age he made copious and critical notes on the textbooks of Kaestner. Later he carefully worked through the great memoirs and treatises of Euler and Lagrange. Yet it was no modern author, it was Eudoxus of

Cnidus, to whom Bolzano owed his decisive Damascus experience, his hour of conversion to the austere beauty of critical and rigorous mathematics. To cheer himself during an indisposition, he narrates, he took up the *Elements* of Euclid, and lit upon the theory of proportion in the Fifth Book, taking so great a delight in its logical depth that the indisposition was swiftly dispelled. Eudoxus there classifies homogeneous couples of numbers or of magnitudes by means of an isoid relation whose reflexivity and symmetry are corollary to the specification, but whose transitivity is anything but trivial. No general theory of isoid relations can be expected at this moment of venture, but the modern algebrist is thrilled to see Eudoxus recognise and fulfil the duty of proving this transitivity, which he does in the deceptively obvious Proposition 11. With the *Paradoxien* in mind, it can be noted that his criterion was not 'entscheidungsdefinit.' The trait in Bolzano which attracted him to Kaestner 'on account of his painstaking demonstration of the obvious' was thus strengthened with time, and bore early fruit in his own proof, dated 1817, that every number lying between two values of a continuous function is itself another value of that function. The sequel will show him in the converse act of disproving something 'obvious' but false, and producing a counterexample earlier by some thirty years than the classical one by Weierstrass.

Bolzano's teachers had been Stanislaus Wydra, a member of the Society of Jesus until its suppression, who had written on analysis and on the history of mathematics in Bohemia and Moravia, and to whom the *Betrachtungen* of 1804 were dedicated, together with Franz Joseph von Gerstner, the founder of the Prague Polytechnicum, who strongly recommended Bolzano for the chair of mathematics on the death of Wydra. Although the solitary rival, Jandera, who had already deputised for several years, was elected in his stead, Bolzano was highly commended for his *Betrachtungen* and for his singular pedagogical talent, and granted prior claim to future appointments. A few months later, however, on 13 February 1805, he was nominated to the new chair of the philosophy of religion, ordained priest on 7 April 1805 by his future protector, Archbishop Chlumčanský, and installed at the University on 19 April. The five mathematical works published during his life-time all belong to the fifteen years during which he held this chair, whereas the *Paradoxien* were written at the close of his life and only published after his death by a friend.

Yet the years of retirement were not years of mathematical idleness. One consolation on the morrow of Bolzano's deposition was the prospect of working out and being able to publish his mathematical ideas; letters to friends tell of plans and changes of plan for various didactical treatises; and papers were read on special themes in mathematics and its methodology. The entire manuscript remains number several thousand sheets, and the scientifically important parts, housed in the Vienna National Library, began to be edited by the Bohemian Society of the Sciences in 1930. Some of their mathematical high-lights had remained unknown until their divulgation by Jašek in 1922.

The chief event in the career of Bolzano was his deposition from the chair of the philosophy of religion at the University of Prague on 24 December 1819, after fifteen years of easy and uneasy tenure. The event has several times been treated rather *a priori* and apographically. Thus, in his textbook of logic 'on a positivistic basis,' Theodor Ziehen lets Bolzano be dismissed for refusing to retract his doctrines (B44, 175) and allows the blame to rest by default with his ecclesiastical superiors. S. H. Thomson's *Czechoslovakia in European History* ventures farther, and stills his voice after 'charges of heresy brought by the Jesuits' (1). The fourteenth edition of the *Britannica* makes space enough for a Bolzano suspended 'from his priestly functions,' though without stating how many functions or on what grounds or whether by a competent authority at all, but makes no space for a single word on Bolzano the mathematician. The suggestion that St. Clement Maria Dvořák (Hofbauer) worked against Bolzano in Vienna was examined in 1931 and found (B5, 419–420) to be baseless.

Recent and exacter research has corrected as well as enlarged the story of this deposition. In a short introduction to Bolzano the mathematician and logician, we must be content to indicate the true context, the authentic sources and the real seat of hostility. The true context is given by the Congress of Karlsbad in August 1819 and the ensuing legislation of 20 September. The authentic sources, the case documents on each side, were transcribed from the Austrian State Archives by H. Fels and published (B5, 419–447) in 1931. The real seat of hostility was the civil government in Vienna.

To begin with, the chair of the philosophy of religion was a new one, which the students themselves resented bitterly at first. Contact with Bolzano, however—the ideal teacher, aglow with love for knowledge and learner alike, patient in starting from the known, meticulously clear in the analysis of terms and reasonings, chary of imposing a great outlay for books, as strict with the titled as with the untitled students and quietly helpful to slender purses—soon converted their distrust into steadfast and wellnigh compromising devotion.

Moreover, the universities under Austrian control had not yet adopted the reforms initiated at Berlin by Wilhelm von Humboldt on the very morrow of Jena. These reforms had stepped out towards ideals perhaps most classicially expressed in Latin by the late Professor of Education in the National University of Éire (2). The universities were to do more than prolong the secondary school or manufacture upwardly meek officials. In Austria, as elsewhere, another type of university persisted awhile. Bolzano would be expected to base his lectures on a standard text: but the chair being new, he was both authorised by the diploma of confirmation (22 October 1806) and encouraged by his friend, Milo Grün, Premonstratensian Abbot and Director of the Philosophical Faculty, to work out a course and submit it after several years of revision and experiment; unfortunately Grün was succeeded on 24 July 1816 by Wilhelm, a man not unfriendly to Bolzano, but timorous. Opposition had gathered, mainly it would seem from Burgpfarrer Frint in Vienna, whose incomplete and expensive text had not been adopted by Bolzano, but criticised as early as 1808. But for the protection of the Archbishop of Prague, he might have been deposed at that time.

Bolzano's written defence (B5, 424–435) dated 31 March 1818 answers the objectors with a wealth of distinction, a completeness of articulation, a relentlessness of dialectic which must have galled minds less analytical, and this the more in that the language cleverly united tact with frankness and dignity with respect. The portion concerned with Frint criticises his work on pedagogical and financial rather than on doctrinal grounds; it intermingles generous recognition of good qualities achieved and severe difficulties overcome; and it deplores the unsettlement of student minds by faults which the talented ones can perceive unaided and uninstigated. This defence offers an interest beyond the mere case history, for it develops ideas on university education which are still pertinent and valuable, and strongly adumbrates the expository principles and ideals which are worked out in detail in the Fifth Part of the *Wissenschaftslehre*.

The routine intermediaries forwarded the defence to Vienna, urged no action against Bolzano and excused some apparent errors as the attempts of genius to explain the inexplicable. The late and only answer to this plea (22 May 1818) was an imperial order (24 December 1819) whose provisions and motives must be briefly noted. One of the duties of the chair was the preaching of exhortations, or university sermons, to the students. Some of Bolzano's views could not fail, at that time, to offend the civil authorities in Vienna. A time would come, the students had been told, when war, that absurd attempt to prove might by right, would be held in the same abhorrence as duelling; a time would come when the myriad differences of rank and social sphere, which wrought such evil, would be reduced to their proper limits, and when each man would treat his neighbour as a brother; and civic obedience, far from being absolute, was limited by conscience and legitimacy. The fate of such pronouncements, and of others on private ownership (3), was at the time of the Karlsbad Decrees a foregone conclusion. The suspicions of the emperor, by whomsoever focused or fomented, were now aroused on their own account. The imperial order, signed 24 December 1819, and communicated to Bolzano on 20 January 1820, laid down: that Bolzano must be dismissed; that he must be reprimanded for the passages on war and rank and civic obedience, whereby he had 'grossly infringed his duties as a priest and a tutor and a citizen'; that as a special favour no punishment would be imposed; that the archiepiscopal curia 'must' examine his works and the notes of his students, and forbid him the pulpit until he gives satisfaction; and the ecclesiastical censor who passed the *Erbauungsreden* must be reported to Vienna (B5, 441–442).

This censor was Dr. Dittrich, professor of church history. When called to book by Vienna, he courageously stood by the passages, considered their misinterpretation sufficiently guarded against, pointed out their moderation compared with mechanical equalitarianism, and disclosed Bolzano's willingness to hear advice at the time. The result was a reprimand for himself and punitive measures which he died too soon actually to suffer.

Under pressure from Vienna, the Consistorial Council pronounced its verdict on 27 May 1820, but in Bolzano's favour, declaring him to be an orthodox Catholic whom it would not be possible to brand as a heretic. When transmitting the verdict, the Archbishop begged that the civil suspension from preaching be abrogated. The emperor's final decision, given on 10 October 1821, renewed the civil suspension and forbade the appointment of Bolzano to any teaching office whatever, and even to any post at the observatory. Troubles continued for other three years, and the Archbishop received on 25 June 1824 an autograph command from the emperor requiring him to demand of Bolzano an express recantation. With every willingness to make the Professio Tridentina, Bolzano was not willing to obey this particular order of the civil authority; so on the last day of 1824, a ceremonial act of acquittal took place in the archiepiscopal palace, and the affair ended amicably on the ecclesiastical level. At one time, soon after 23 May 1822, the Archbishop had already confided to Bolzano that he himself was under pressure. Vienna at least gave Bolzano a pension, which was afterwards tripled at the instance of Count Leo Thun.

For fuller accounts of the deposition and its backgrounds, the reader can only be referred to Chapter IV in Winter (B8) and to the documented monograph by Fels (B5). It is already clear, however, that some accounts are too simplified. At various stages in the affair Bolzano had been defended by more than one archbishop, more than one abbot, more than one professor of theology, by the Consistorial Council, occasionally by the regional government and towards the end even by lay and ecclesiastical members of the Court Study Commission itself.

Though wounded by this separation from academic youth and by this rebuff to his educational ideals, Bolzano was not daunted. The last twenty-eight years of his life were spent in retirement, writing on a considerable range of subjects—logic and mathematics, aesthetics, philosophy and religion, linguistics and biography, sociology and physics—and publishing his works as best he could under the state censorship imposed at Karlsbad and meant to control publication even outside Austria. They consequently set a number of bibliographical puzzles, though not in the fields of logic and mathematics. From 1815 onwards Bolzano was an active ordinary member of the Bohemian Society of the Sciences, whose *Berichte* and *Abhandlungen* yield a total of thirty-four readings and seven published memoirs from his pen. The twenty summers from 1823–1842 were spent on the estate of his benefactress, Frau Hoffmann, at Těchobus near Pacov (Patzau) and twenty-two kilometres north-east of Tabor, and the winters at Prague with his sole surviving brother. After the death of Frau Hoffmann in 1842, Bolzano returned to live continuously in Prague save for visits to such friends as Anton Veit at Liběchov (Liboch) just below the confluence at Mělník of the Vltava (Moldau) and the Labe (Elbe), and here it was that he discussed the *Paradoxien* and other works with their posthumous editor, Přihonský. He died at ten in the forenoon of 18 December 1848, and was buried three days later in the Wolschaner Friedhof. Even the timorous Wilhelm, reporting Bolzano frankly 'for his own safety's sake as an official' on 5 May and 26 October 1817, did not shrink from acknowledging Bolzano's 'untiring diligence and zeal, deep thought and insight, comprehensive learning, outstanding competence and unremitting care to render the students, by teaching and example, meritorious as citizens, and useful to mankind' (B6, 439).

Fesl's choice of a motto for the Autobiography (B6) from Xenophon's eulogy of Socrates in the *Memorabilia* (I, ii, §61) was singularly merited by Bolzano: the teacher inspired and inspiring and one of the pioneers in the reconquest of logic in mathematics.

Observation

Although the Consistorial Council in Prague absolved Bolzano from heresy in the theological sense and in the matters raised against him, it must be placed on record that some of his philosophical theses did lie outside the scholastic traditions. Thus, his opinions that all created substances exist in space and time, that in all God's creation there is no lowest and no highest degree of being, and that eternity is the same as time stretching to infinity in both directions (*Paradoxien*, §17, §39, §55, §58) would not be accepted by Catholic philosophers and theologians, and must be set down as opinions personal to Bolzano. In this work they are incidental rather than systematic. Interested readers are therefore referred to the careful studies by E. Winter and H. Fels. The former gives (B8, Chapter 1) the historical background to these divergencies in an author of Bolzano's position; the latter, writing under scholastic auspices, gives (B59) a sympathetic but critical account of Bolzano's philosophy as a whole.

NOTES TO II: BIOGRAPHICAL SKETCH

(1) Princeton 1943, p. 127.

(2) T. Corcoran S.J., *Renovatio litterarum in scholas saeculi a Christo sextodecimo deducta.* Ad usum academicum, Baile Átha Cliath, 1925. P. ix, lines 16–24 translate as follows: 'The influence of a teacher will be at its height when he propounds sundry norms of judgement and stimulates his hearers to exercise their own skill and ingenuity in forming a truly personal conclusion based on original sources; having digested and elaborated it, to discuss it freely and severely; having discussed and revised it thoroughly, to express it clearly and concisely, to illustrate and corroborate it from the writings of others; and finally to adapt it with care and diligence to the needs of our own time.'

(3) In the dissertation *Vom besten Staate* (B34 and 35) Bolzano accepted private ownership in principle, but within severe limits. The state might regulate its exercise to meet the exigencies of the common good, and non-use as well as ill-use could cause forfeiture of property rights. The hoarding of rare manuscripts and art treasures was heavily censured. The franchise was restricted to such as knew the issues and were affected by them, but civic instruction was in time to become universal.

(4) For a full English translation of the Karlsbad Decrees on the state control of universities and periodicals, see the *Introduction to Contemporary Civilisation in the West.* New York (Columbia University Press) 1 (1946), 124–130.

III. Bolzano the Mathematician

HAVING seen mathematics recover dialectical discipline, turn a little less to quantity and a little more to relation, deflect a portion of interest from physical applications to logical analysis and thus win new ideas which physical science would gladly come to employ for its own ends, having observed that Bolzano's life links the two periods in time, and having outlined the general course of that life, we must examine the content of his mathematical writings, acknowledging the survivals amid its new arrivals, and the retrogressions amid its progress. Taken alone, the *Paradoxien* would give a seriously unrepresentative picture of Bolzano the mathematician, and this all the more in that we have cause to suspect its declared and undeclared editors of pandering to the mathematical Mrs. Grundy of the day. We must certainly draw upon the other five short treatises composed between 1804 and 1817, when Bolzano was still in the chair of religious philosophy, upon his logical master piece the *Wissenschaftslehre* of 1837 with its wealth of mathematical paradigm, upon the papers read to the Bohemian Society of the Sciences, upon mathematical remains but recently published or described and not least upon works of Bolzano not at all obviously mathematical. His mind was singularly integral as well as versatile, and we must be prepared to encounter passages of considerable importance for logistics or pure mathematics in such unlikely places as the dissertations on the composition of forces and the tridimensionality of space.

Separating then Bolzano's mathematical doxography from his mathematical material, we shall begin with his views on the nature and method of mathematics, and in particular with his answers to questions of definition and existence. Then, after watching his treatment of special topics such as holomerism, the theory of sets, the theory of real functions and the arithmetisation of analysis, we shall pass in review a number of characteristic excellences together with a few points which, even after correction for historical parallax, still tend to disappoint the modern reader by their contrast with those very excellences. Finally, we shall glance at Bolzano's reception by the mathematical world, and give some appreciations.*

A. DOXOGRAPHY

1. *The Nature of Mathematics*

Cogency and definiteness place mathematics nearest to the ideal type of science (BD, beginning) and hide its austere perfections from many who are sensitive to other forms of beauty (BS, 25–26). None can be good philosophers who are not first good mathematicians, and Hegel's confessed bewilderment over simple mathematical proofs augurs ill for his philosophy and for its followers; in fact, the abstractions of philosophy are infinitely more difficult than are those of mathematics (*ibidem*).

Bolzano changed with time his views on the distinctive object of mathematical science. In the prime of his life he regarded it as a formal science whose every truth is 'purely conceptual,' and which is busied more with relation than with quantity. In the long album of definitions he then made an entry very similar to that of Benjamin Peirce, declaring mathematics to be 'the science of the general laws according to which the existence of all things is regulated' (BD, 11) and 'all things' is to include 'our intuitions, our universal ideas and whatsoever has objective existence independently of our consciousness.' As an old man, on the other hand, Bolzano withdrew this thesis, saying (EK, 135) that he had since reconciled himself to the commoner opinion alleging it to be the science of quantity, and that he was not ashamed to make open recantation. Our study of Bolzano the physicist will show in the sequel, however, that in an exactly coaeval investigation of certain quantities, he has by no means abated his eager search for their logical structure

* To make incidental documentation available yet unobtrusive, the abbreviations expanded under the table of contents will henceforward be freely employed.

above all else; and in this very investigation he reaffirms the 'purely conceptual character' of all mathematical theorems and their dissociation from 'intuition.' His repentance is therefore more likely to have borne upon the trespass into metaphysics than upon any trespass into logic pure and simple.

2. *Mathematical Method*

Like De Morgan, his partial counterpart in England, Bolzano insisted strongly on the need for a mathematical methodology, and for more collaboration between logic and mathematics. He complains that so few of his contemporaries are both mathematicians and philosophers, and is anxious in all his mathematical writings to be judged by these few alone (DD, 203). His delight would have been great indeed could he have foreseen our modern fundamental theories.

Ancient axiomatic ideals are reviving, modern ones are foreshadowed when Bolzano censures current mathematical literature for taking so little trouble to ascertain which theorems are primitive in a context and which are derivative (WL, §577). Proofs which establish points of this nature should be grouped together under a separate name. Those sorry persons who are content with mathematical facts and indolently neglect their objective grounds are disdainfully rebuked as wanting the very appetite for scientific knowledge (ZK, 14). Some of the axiomatic analysis is used at once in philosophy, for example: the proposition $7+5=12$ can be deduced (1) from the associativity of addition and catenary definitions of the natural numbers (WL, §315, 186) and is, therefore, in the senses carefully formulated by Bolzano, not 'synthetic' but 'analytic' and, in particular, free from 'intuitions.'

Logical impeccability and definiteness in the premisses are not enough to render a proof quite satisfactory. Bolzano speaks out again and again, in work after work, against proofs which barely and adventitiously succeed, calling always for others which, in addition to being valid, also disclose the 'objective grounds' why a proposition be true, and the precise locus it occupies within the surrounding theory. We must recur to this principle, which in the eyes of Bolzano extends far beyond mathematics, when we consider him as a logician: but we ought to note in passing that he carries it farther than we relish. He objects, for example, to the use of circumscribed and inscribed polygons in the rectification and complanation of curves (DP, introduction) on the score that they be a foreign element, 'adventitious and subsidiary' relative to a curve proper. But we must ask how else we could do so much as define arcs and areas, and how else distinguish the cases where arcs and areas exist from those where they do not exist (2).

Relation structure is worth exploring for its own sake, but also as a powerful instrument in the unification of mathematical problems and theories. An abstract and all but logistic definition of 'to determine' and 'similar to' is laid down (ZK, §§4–6) and welcomed, alongside a correspondingly abstract 'principle of similarity' (ZK, §7) as helping simultaneously to solve mathematical questions which are materially diverse but structurally identical. Certain simple measure functions are sought (PU, §40) by the direct intervention of an isomorphism between the addition of numbers and the juxtaposition of lengths, areas and volumes. In yet another memoir, a problem in statics is reduced openly (ZK, §33) by isomorphism to a problem in geometry and ultimately to one in analysis. In short, our Bolzano is commencing the search for pervasive relation-theoretical principles as well as for accuracy in logical detail. One of his ideals (DD, 213) is the kind of principle whereby we infer that an object N verifies the predicates $n, n', n'', \ldots$ from the two premisses (i) that a given object M verifies the predicates $m, m', m'', \ldots$ and (ii) that a given relation subsists between M and N. Attempts at principles of so abstract a type are a refreshing change from the annihilation of mathematically naughty particles. Bolzano himself gives to this type the name 'higher rule of three.'

These essays in mathematical methodology were no mere velleity. Bolzano made strenuous efforts to carry out his own principles, and to expound them systematically. This is done most professedly in the *Wissenschaftslehre* of 1837, which contains a scattering of little subtreatises on the pedagogy of proof (§§512–537) or the grammar and style of symbolism (§§637–698) or the very art of writing treatises (§§392–718). In connexion with the *Paradoxien* it should be mentioned that the paper read to the Bohemian Society of the Sciences in February–June of 1842 on '*The particular march of ideas to be followed when setting forth mathematics in a rigorously scientific manner*' had its third instalment devoted to 'the important doctrine of the finite and the infinite.'

3. *Scrupulousness in Definition*

Even when Bolzano defines a term less precisely than is possible to us beneficiaries of his reforms, as happens for example with the term 'continuum' (PU, §38), we still can sense how very conscientiously he goes about the task of framing an exact definition. Illustrations of his carefulness abound so greatly, though not always at their best in the *Paradoxien*, that we may turn at once to his stated canons.

First of all, he was solicitous that definitions should be essential and not merely true by concomitance. The rectification integrand for plane curves is not to be taken blindfold but deduced (DP, at beginning) from some previous and genuine definition of arcual length, just as 'freedom from jumps' was to follow from a previous and genuine definition of a continuous function, and not be an easy-going substitute for a definition. These maxims, excellent in themselves, were also carried rather far, as when Bolzano sets out to 'demonstrate' (DP, §10) that when $f(t)$ represents the path of a particle, then $f'(t)$ and $f''(t)$ will represent its velocity and acceleration.

Bolzano would have disapproved of the schoolbooks which issue specifications, say of the function $f(x, y)=x^y$, in uncoordinated instalments, patched up after frantic hunts for 'meanings,' with no care to subsume previous partial definitions by the principle of isomorphic subdomains, and no care to leave matters so that higher stories can be built on, if they must be built later, without ugly cracks or despairing demolitions. No, definitions were to be given whole (WL, §79, 372) and not by patchwork—or at the very least, subsequent enlargements must be made organically. Here again Bolzano practises what he preaches, even in statics. When integer multiples of a force have been defined, he warns us, its fractional and irrational multiples still remain to be appropriately defined (ZK, 5). The same rule obtains in logic too: after incompatibility has been defined for non-empty classes, Bolzano extends his definition to empty ones by 'variation of residual arguments' (WL, §108).

Lastly, Bolzano separated essential from nominal definitions and gave to the latter the peculiar name of 'willkürliche Sätze' or 'arbitrary positings.' More momentously still, he demanded that the act of defining and the act of establishing existence should be performed as two clearly distinct acts, and not vaguely merged into one.

4. *Scrupulousness concerning Existence*

Trains of mathematical thought can very well begin on psychological occasions very different from the epistemological character of the theories in which they issue. The occasion has been provided at various times by urgent physical problems, the proportion of stimulus from this quarter being exaggerated by some at the expense of the stimulus from ideals of logical exactitude and explicitness, or from the artistic spell of the theory of numbers. Be this as it may, errands run for the natural scientist bring to the mathematician losses as well as gains, and one of the losses, having been fought against by Bolzano, must be noticed here. The background of a physical inquiry, which nature herself seems to answer in one way and only in one way, is liable to dull our sense of the problematic character of existence and of uniqueness. At one level, we have the Thomson-Dirichlet Principle and its sanation by Weierstrass (3) and Hilbert; at another level, we have the difficulty, so long as derivates are too closely associated with velocities, of conceiving continuous undifferentiable functions, of accepting them when they are constructed, and of profiting from their stimulus to real function theory.

One mistake of the kind was to suppose that when every entity in a class is associated with a number, then the class of associated numbers must, if bounded above or below, contain a maximum or minimum member. Bolzano exposes this error clearly and emphatically. After admitting the existence for finite sets, he points out the possible non-existence for infinite ones (PU, §62), asserts instead the existence of an upper or lower (strict) bound which may turn out to be a non-member of the set (RB, §12), draws out the contrast between bounds and extrema (RB, §13) and insists on the importance of distinguishing them in geometry and mechanics as well as in analysis. Two other paragraphs of the *Paradoxien* (§32 and §47) contain between them no fewer than four precautionary existence hypotheses. The provisional definition of 'infinite' sets, prior (§9) to the introduction of holomerism as such, is guarded by the promise of an existence proof to come, and that existence proof —criticised though it would come to be by Russell and Fraenkel —was in §13 duly offered.

To collect more examples of this existential rigour would encroach upon the study of Bolzano's specific topics. We may content ourselves, therefore, with an example of his punctiliousness in the matter of uniqueness. Seeking a function apt to express the equilibrium condition for a set of forces, and listing the properties such a function must have, Bolzano notes that a certain elementary function does have those properties. Nevertheless, he is not to be deflected from the path of logical duty: he lays down that duty in set terms (ZK, §52) and spends the rest of his space in trying to fulfil it, thus grasping the fact that the duty of proving a mathematical uniqueness theorem is not abrogated by an accidental uniqueness in some physical application.

B. SPECIFIC TOPICS

1. *Holomerism*

Bolzano's *Paradoxien* are quoted far and away most often for their recognition of a phenomenon to which we shall give the short name of 'holomerism.' It consists in this: that a set S of entities contains a proper subset s such that, by means of a fixed relation, each member of S corresponds to just one of s, and by means of the inverse relation, each member of s corresponds to just one of S. In words: the main set is in biunivocal correspondence with one of its proper subsets. Three historically interesting cases are: (i) let S comprise all the natural numbers, and s the squares among them, the relation linking n in S with n^2 in s; (ii) let s comprise the real numbers x such that $0<x<5$, and S the real numbers y such that $0<y<12$, the relation being specified by $12x=5y$; (iii) let S comprise all the points in a human body, and s all those in a certain finger, the correspondence being effected by a suitable geometrical transformation. The appearance of having one infinite set paradoxically richer in members than another infinite set is also produced when we compare (iv) the set of all the diameters of a fixed circle with the 'twice as numerous' set of all the semicircles into which those diameters divide it.

Our Bolzano does more than bring two curious and isolated examples of holomerism, namely (ii) and a geometrical analogue. He asserts its impossibility with all finite sets (PU, §22) and its possibility with all infinite ones (PU, §20) and thus erects it into a criterion of infinitude—though without incorporating it into a formal definition of infinitude in sets. As regards the mere discovery of cases, he had been anticipated several times.

Galilei is well known to have tendered (i) in the *Discorsi* (4) as early as 1638. He also noted the additional strangeness that despite the holomerism, the subset members occur in the natural sequence with increasing rarity among the set members. Probably less familiar are the anticipations of (iii) by Plutarch (5) and of (iv) by Proclus (6). Antiquity has even bequeathed to us the spectacle of a layman's resentful bewilderment at the phenomenon, on the part of Philoponus (7) in his counterblast to Proclus' *De aeternitate mundi*.

The anticipation by Galilei offers a point of fine irony. The *Discorsi* contain two contiguous passages (8), an overt pseudo-proof that the periphery of a circle be of no greater length than the central point, followed by example (i) of holomerism. The former passage is actually quoted by Bolzano (PU, §46) although the latter passage, so much more apposite to the *Paradoxien*, is entirely overlooked! The reason is not far to seek. We are frankly informed that Bolzano took the pseudo-proof at second hand from Kaestner.

Our first sign that the *Paradoxien* are under way takes us back to the year 1842. As late as 1837, on the other hand, Bolzano seems not yet to have lit upon the idea of holomerism, for his treatise on logic (WL, §87) simply states that two multitudes can be paired off either exactly, or until an overplus of members is left in one of them; it distinguishes finite and infinite sets only in the provisional manner of *Paradoxien* §9, and rejects infinite number outright. In fact, as soon as we have said that Bolzano formulated holomerism more precisely than hitherto, viewed it more universally, and made it a symptom of infinitude in sets, we have claimed as much as the *Paradoxien* warrant. Neither there nor in other writings does Bolzano exploit holomerism enough to be regarded as a precursor of Cantor in any greater measure than that. Indeed, it must be regretfully confessed that the *Paradoxien* themselves, as they stand, contain several apparent backslidings.

Thus, we for whom Cantor has won the idea of power or cardinal number are sorry to read in §18 that a series has fewer terms after decapitation than before, in §39 that a short duration contains fewer instants than a long one, in §40 the argument that, since a cube has the eightfold volume of a cubic octant and since the inner octant surfaces are duplicated, volumes cannot be proportional to the number of contained points, or in §41 that the number of points in similar figures is in proportion to their sizes, and in §42 that J. K. Fischer errs in supposing the numbers of those points to be the same, or finally and drastically in §48 that volumes contain more points than their bounding surfaces, and surfaces more points than their bounding lines. Has not Cantor proved (9) the contrary and more, and this precisely in terms of biunivocal correspondence? Could we not fairly expect Bolzano to apply his own criterion from §20, even if faultily, or without the ingenuity of a Cantor or a Peano?

No, we cannot fairly do so. Nor can we blame Bolzano for these passages, because §21 makes it clear that, while *holomerism* is recognised as liable to occur, and even bound to occur within individual infinite sets, *biunivocal correspondence* between two sets is not being erected into a criterion of their equinumerosity. It is only contended—and thus far rightly, thus far salutarily, thus far in advance of the time—that biunivocal correspondence does not preclude the case where one aggregate is identical or equipotent with some main set, and the other aggregate identical or equipotent with a proper subset of that main set. Still less are we to look for the conception of biunivocal correspondence as an isoid relation partitioning multitudes into classes with the several ones of which the cardinal numbers will afterwards and for a time without challenge be severally identified.

For these things we must wait till Cantor and Frege; but in the meantime holomerism had once at least been recognised afresh, and once more by Bolzano's English counterpart, De Morgan (10), who saw it holding between the base of a triangle and a transversal in perspective from the vertex. De Morgan observed also that the half of a rectilinear segment could be put both into one-one and into two-one correspondence with the whole segment.

2. *Aggregates*

In §§3–5 of the *Paradoxien*, borrowed from §§82–84 of the *Wissenschaftslehre*, Bolzano lays down what follows: (i) an 'Inbegriff' is a collection of members with definite order permitted but not enjoined, the underlying conjunction being treated as a logistic function; (ii) a 'Menge' (set) is an 'Inbegriff' in so far as invariant under permutation of members; (iii) a 'Vielheit' (multitude) is a 'Menge' whose members belong to a single logical species; and (iv) a 'Summe' (sum) is a collection in so far as invariant under the dissolution of subcollections into their own separate members, comparison being made with the associativity of addition. For a modern criticism of these definitions and of Bolzano's existence proof for infinite sets by means of hierarchical superpredication, see both Russell's *Principles of Mathematics* (11) and his *Introduction to Mathematical Philosophy* (12). Curiously enough, while insisting that infinite sets exist, Bolzano remains doubtful concerning sets with no member or only one member, and accounts himself generous (WL, §84) in allowing 'as few as two members'!

The subsequent deepening of these problems makes it worth remarking how urgently Bolzano demanded that infinite sets be, in the terminology of Oskar Becker (13), 'entscheidungsdefinit.' The infinitude of a set must not (WL, §87, 410) be construed as indefinitude, or as 'not being given as a whole.' With infinite sets, we do not have to wait until the members are strewn about before us: it is enough that we can decide concerning any one proposed entity whether it belong to the set or no (WL, §87, 412 and PU, §26 *ad finem*). Two such sets are identical *qua* infinite sets only when their specifications are identical (PU, §24).

§101 of the *Wissenschaftslehre* raises in its caption the arresting question: whether every arbitrary set of objects be accompanied by a universal concept embracing precisely those objects? Bolzano grants this in the avowedly trivial sense that a disjunctive predicate can always be constructed *ad hoc*, but in any sense less trivial, he hastens to add, the answer is very doubtful.

3. *Real Function Theory*

Two major achievements stand to the credit of Bolzano in the theory of continuous functions of a single real variable, one of which has long been known as his, and the other of which only came to light in 1922.

The occasion of the former and better known achievement was one of Gauss' four proofs (14) of the fundamental theorem on rational integer functions (15) that every such function has in the complex domain exactly as many zeros as its degree indicates. Gauss himself was aware of a gap due to the intrusion of geometrical intuition into a purely analytical matter, and planned to fill the gap at a future moment. Bolzano perceived the same gap, and also planned to fill it (RB, 3), but, while deliberately arithmetising, he plainly disavows purism for purism's sake (RB, 10) and is only anxious lest examples be smuggled in as a substitute for genuine proof.

His first care is to purify the definition of continuity. The property of having as a third value any number lying between two values, accrued indeed to continuous functions, and would be duly demonstrated: but that property was inept to serve as the formal definition. Some more primitive property must be sought, and the above one punctiliously deduced from it. The reformed definition of continuity, as a point x in the interval for which it is claimed, runs thus: 'that the difference $f(x+\omega)-f(x)$ can be made smaller than any preselected quantity once we are permitted to take ω as small as we please.' The difference is tacitly taken in absolute value, and the conduct of the proofs reveals that the ideas are somewhat more precise than the wording given. In paraphrase therefore: $f(x)$ is continuous over $a<x<b$ if, howsoever we select ξ with $a<\xi<b$ and ε with $\varepsilon>0$, there exists δ, more explicitly $\delta(\varepsilon, \xi)>0$, such that $|t|<\delta$ entails $|f(\xi+t)-f(\xi)|<\varepsilon$.

Having thus cleared the ground, Bolzano enunciates (RB, §15) the standard theorem that if $f(x)$ and $g(x)$ are both continuous over $a<x<b$, and if $f(a)<g(a)$ but $f(b)>g(b)$, then ξ exists such that simultaneously $a<\xi<b$ and $f(\xi)=g(\xi)$. Incidentally, we may instructively compare with this a passage in Proclus (16) where certain non-Archimedean magnitudes are discussed, and it is stated that in their case the passage from greater to less does not lead through equality. The proof depends essentially on another theorem to which we shall revert before long.

Meanwhile, with the aid of his improved definition, Bolzano is able to enrich the theory with a number of algorithmic propositions. The difference of two continuous functions is continuous (DP, §2) and so is a continuous function of a continuous function (DP, §3).

Let us now revert to the auxiliary theorem in §12 of the *Rein analytischer Beweis*. The enunciation is adapted to current purposes and disguises the modern form of the theorem somewhat. A certain property M is under discussion, and the actual assertion can be analysed without paraphrase as follows: *Hypothesis* (i): not every x has M. *Hypothesis* (ii): $x<u$ entails: x has M. *Conclusion*: there exists a number U as greatest among the numbers of which it can be said that every predecessor has M. This conclusion combines two statements. We resolve it thus: *Conclusion* (i): every predecessor of this U does have M. *Conclusion* (ii): whenever $U'>U$, then not every predecessor of U' has M.

The whole enunciation can now be paraphrased by writing $\mathfrak{M}$ for the class of numbers which does not possess M. It then runs: *Hypothesis* (i): $\mathfrak{M}$ is not empty. *Hypothesis* (ii): $\mathfrak{M}$ has no members below u. *Conclusion* (i): U exceeds no single $\mathfrak{M}$. *Conclusion* (ii): whenever $U'>U$, then U' exceeds at least one $\mathfrak{M}$. In modern terminology: A non-empty and lower-bounded set of numbers possesses a 'finis inferior.'

The proof is by successive dimidiation, but of left halves only and recommencing at different places. Compressing notation and abbreviating steps, let us write for our own convenience

$$D_m = D/2^{k_m}$$

where D is positive and high enough for $u+D$ to be one of the hypothetically existent numbers exceeding a non-possessor of

M, and (k_m) is a monotone rising sequence of positive integers so constructed by Bolzano that

$$u_m = u + D_1 + D_2 + \ . \ . \ . \ + D_m$$

is preceded exclusively by possessors of M, whereas

$$v_m = u_m + D_m$$

is preceded by at least one non-possessor. At each stage Bolzano either discerns in the fresh u_m the asserted U, or else continues. If he must always continue, he generates a sequence (u_m) which fulfils a previously demonstrated (§7) convergence condition, and whose limit U is then shown to verify the original assertion.

We are thus driven farther back still to §7, and discover there the 'two term' convergence criterion for sequences, or the remote segment criterion for series, published some years before the *Cours d'Analyse* of Cauchy, the trivial necessity occupying §6 and the non-trivial sufficiency §7. Bolzano asserts that the sequence (F_n) has a limit, say X, whenever, $\varepsilon > 0$ being preselected, we have $\nu = \nu(\varepsilon)$ such that $p > q > \nu$ entails $|F_p - F_q| < \varepsilon$. Unfortunately, the strict existence proof is marred by a fallacy. Bolzano sets out to do two things: to show in the first place that the existence of a limit X is not self-contradictory, and in the second place that this limit X can be approximated as closely as we desire. Now the latter enterprise is conducted aright for all the cases where a limit does exist, and as follows. Preselect $\delta > 0$ and $\varepsilon > 0$, whereupon (i) in virtue of the two-term condition, $|F_{n+r} - F_n| < \delta$ for suitable constant n and almost all r (that is, with at most finitely many exceptional r) and (ii) in virtue of X being the limit, $|X - F_{n+r}| < \varepsilon$ for almost all r, but with r eliminating itself and n constant, more simply $|X - F_n| < \delta + \varepsilon$. Since ε was positive but otherwise arbitrary, it follows that, more curtly, $|X - F_n| \leqq \delta$. Consequently, it is possible so to choose the particular term F_n that it furnishes the desired approximation to X. But surely—this would follow, would it not, quite directly from X being the limit? Why the intervention of r at all?

The trouble on the surface is, naturally, that the absence of self-contradiction and the possibility of approximation once existence is secured are together insufficient to *establish* existence. The trouble below the surface is, just as naturally, that in

1817 no one, not even Bolzano, had constructed adequately wide categories of real number—the phrase 'all real numbers' being advisedly avoided. Biding our time for the constructions by Kronecker and Weierstrass and Dedekind and for the challenges by Brouwer and Weyl, we must applaud Bolzano for bringing these two theorems to the light of day at all, and for appreciating (RB, §13) their great importance and wide scope in mathematics, geometry and mechanics.

The Pre-Weierstrassian Counterexample

After catching Bolzano in a moment of weakness, we can happily end by catching him in a moment of unusual strength. On 16 December 1921 M. Jašek communicated to the Bohemian Society of the Sciences in Prague and on 20 September 1922 to the Deutsche Mathematikervereinigung in Leipzig some results of a more diligent examination of Bolzano's mathematical manuscripts. His principal discovery was that Bolzano had invented, by 1830 at the latest, a process for the construction of *continuous but undifferentiable functions* (21).

The prevailing belief had been that, apart from innocuously few isolated points, every continuous function is everywhere differentiable. Proofs were even published: in 1806 by M. A. Ampère (17) and as late as 1872 by P. Gilbert (18). But a conclusive counter-example by Weierstrass circulated orally in Berlin from 1861 onwards (19) and was published (20) in 1875. Bolzano's own counter-example, which thus antedated Weierstrass by some thirty years, was contained in his manuscript *Funktionenlehre*, first printed in 1930. The only rival claimant, a function attributed to Cellerier, goes back at the best only to 1860.

The relevant quires of the manuscript are numbered 13 to 19. Quire 16 is noteworthy if only for the independent definition of forward and rearward derivates, and the signalisation of their equality as something which might or might not happen. Du Bois-Reymond was thus not the first to descry this possibility. The three duties of one who proposes a counter-example are stated with dry precision in Quire 13. They are: (i) to specify the function, say $B(x)$, over $a \leqq x \leqq b$; (ii) to prove $B(x)$ continuous there; and (iii) to prove $B'(x)$ non-existent there. Bolzano himself only claims its non-existence at a dense set of points; but as a matter of fact, $B'(x)$ exists nowhere, and even Bolzano's set is dense everywhere.

The three duties are then carried out in order. Quire 17 draws the moral that continuity is necessary for differentiability, but not sufficient. Soundings are even taken into the behaviour of a continuous function at a point of undifferentiability, with the outcome for example that either the difference quotients tend to determinately signed infinity, or else at least one sequence of difference quotients tends to a single finite limit.

The counter-example mechanism is actually simpler in Bolzano than in Weierstrass and resembles that due to van der Waerden. Let (a,p) be an arbitrary 'Cartesian' point and let (b,q) be another with the sole proviso $a \neq b$ and $p \neq q$. Let the phrase 'fundamental operation' denote temporarily the transition from a given single rectilinear segment to a certain coterminous chain of four such segments as follows: when the initial segment joins (a,p) to (b,q), then shall the three elbows of their zigzag join lie at $\left(\frac{5a+3b}{8},\frac{3p+5q}{8}\right)$ and $\left(\frac{a+b}{2},\frac{p+q}{2}\right)$ and $\left(\frac{a+7b}{8},\frac{-p+9q}{8}\right)$. Next, define the nullth function $B_0(x)$ of a forthcoming sequence of functions as linear over $a \leqq x \leqq b$ with end values $B_0(a)=p$ and $B_0(b)=q$. Then define the first function $B_1(x)$ as linear in segments, four in number, corresponding to the outcome of the fundamental operation on the original join; and generally: define $B_n(x)$ as linear in segments, 4^n in number, corresponding to the outcome of the fundamental operation on each of the 4^{n-1} segments of $B_{n-1}(x)$. This sequence converges uniformly to a continuous limit function $B(x)$. And whenever an elbow arises under one performance of the fundamental operation, it remains an elbow under all subsequent ones and yields a point of undifferentiability.

For the circumstances of the discovery of this early counter-example, see two accounts by Jašek, one summary (B49) and one in detail (B51). For a fuller mathematical description, see the latter memoir and two others by G. Kowalewski, one in the *Leipziger Berichte* (B53) and one in the *Acta Mathematica* (B54). For the disclosure of further idiosyncrasies of $B(x)$, see K. Rychlík (B55) and V. Jarník (B48). Among these are: (i) that $B(x)$ has no finite derivate at any point of the interval, and none at its terminals, and no determinately signed infinite derivate at any internal point; (ii) that there is never a coincidence between upper and lower Dini derivates, either forward or rearward; (iii) that a denumerable set of points exists where all four Dini derivates exist and are pairwise unequal; (iv) that a set of points of continuum power exists where the four Dini derivates have respectively signed infinite improper values. There is a version of the longer account by Jašek in Czech (B50).

4. *Arithmetisation*

Bolzano's 'horror geometriae' (to quote the strong expression of Jašek) was certainly no denigration of geometry as a science in its own right, but only a disinclination to let it be applied in patches to hide the weak spots in analysis. He gave fair warning (PU, §36 *ad finem*) that continuity is an unreliable guide, and tried like Gauss to fight intuitional conversatism—more ruthlessly in fact than du Bois-Reymond on the day when the latter unguardedly asked (22) how we were to 'reconcile the results of research in analysis with our native conceptions of quantity'! The *Zusammensetzung der Kräfte* defines conceptual propositions as those in which no single component is intuitional, and subjoins the remark that to this category belong all the theorems of mathematics (ZK, §2).

In part agreement with Aristotle, Bolzano condemns the passage 'from one genus to another,' and in particular the proving of arithmetical or algebraical or analytical theorems by considerations drawn from geometry or mechanics (WL, §536, 297–298) but he takes exception to the example in *Analytica posteriora* (75 a 38–39) since the catabasis, while forbidden in the direction from arithmetic to geometry, is permitted and sometimes necessary in the direction from geometry to arithmetic. Bolzano thus ranges himself at the side of an illustrious early Greek mathematician who maintained the superior clarity and certainty of arithmetic and the need to draw upon it for the completion of geometrical proofs—Archytas (23). For all that, Bolzano warns against, and for his part repudiates, any exaggerated purism (RB, 10) and acknowledges the heuristic value of pursuing interbranch analogies (WL, §536, 298).

The old-time rivalry between analysts and geometers for the palm of rigour and certitude, of explicitness and full axiomatisation, of accuracy as conditioned by subject matter, has of course receded more and more since group-theoretical, ideal-theoretical and similar ideas began to cause the structural features of the two branches to be more similar and more important than the mere constituents in either.

No, Bolzano had a strictly methodological motive at the time. He feared lest the appeal to a foreign subject lead to unconscious sacrifice of generality and render us content with grounds less ultimate than they might be (ZK, §51). Felix Klein justly hailed him (24) as a 'father of arithmetisation.' Himself distinguished alike for mathematical researches and organisation and historiography, Klein considers that Bolzano was stimulated to his investigations in part by the scholastic tradition, with which his priestly training would have made him familiar. With the literature of that tradition, it will be recalled, Cantor himself was also thoroughly acquainted.

C. RECEPTION AND APPRAISAL

1. *Minor excellences*

The bulk of Bolzano's mathematical work having been dealt with, we shall now collect a number of isolated excellences and disappointments from his algebra, from his elementary analysis and from his inchoate real function theory.

FROM ALGEBRA

1. Addition is defined by a catenary specification of natural numbers together with associativity: to this extent just as in Landau (25) but without the later and subtle refinements of Grandjot and Kalmár: once formally (BD, 147) and once informally (PU, §5).

2. The equations $A-A=0$ and $A+0=0$ and $A-0=0$ are not advanced as if sufficient to create a number zero, but only as norms for an eventual definition: and warning is given that the functions $x+y$ and $x-y$ then acquire a supplementary specification. This foreshadows (PU, §34) the explicit transition from

the sum-analogue function of a ring r to the different sum-analogue function of a ring R' having a subring r' isomorph to r, a transition essential to the rigorous construction of successive classes of number. Unfortunately, zero is not recognised as a number in its own right, nor is this to be expected until the advent of modern algebra.

FROM ELEMENTARY ANALYSIS

3. When defining limits, Bolzano demands that a preselected degree of approximation be both attained and retained (PU, §37, 66) and so sponsored this useful pair of participles.

4. We are not only admonished never to divide by zero, but also admonished to take care lest we do so unwittingly during an elaborate calculation (PU, §34). Such fictions as a 'true value of 0/0' are rejected (PU, §36) and a distinction made between values and limits. Unfortunately, Bolzano does it on a conception of zero which modern algebra does not recognise.

5. The definition of a sequence is defended (WL, §85 and PU, §7) against cavils based on iterating sequences, and sequences in which a new term is specified as a given function of several earlier ones. The sequence (a_n) with each term $a_n=1$ is defended against other cavils on the score that, after all, the n-th term is still a perfectly definite function of n.

FROM REAL FUNCTION THEORY

6. We have (RB, 12) a curious if fragmentary and accidental anticipation of Landau's o-notation.

7. Elsewhere (PU, §49, article 1) we find an equally curious foreuse of Hankel's iterated concentrations.

8. At a third point (PU, §48) we have a kind of ordinate set, consisting of normals to the plane at points in a plane curve. The area of the resulting quasi-cylindrical surface is not, however, conceived on the lines of modern integration theory.

9. Nevertheless, lengths and areas and volumes are defined (PU, §40 *ad finem*, and DP, §19) in strikingly abstract fashion as numbers so associated with lines and surfaces and solids that juxtaposition is isomorph with addition. It will be recalled that $m(A+B)=m(A)+m(B)$ enters into Carathéodory's axiomatisation of measure functions.

10. Finally, Bolzano had altogether advanced ideas on the nature of variability both in mathematics and in logistics. This appears in part from a passage (PU, §12, article 1) where he says that a variable quantity is not strictly 'a' quantity at all, not (as it were) a quantity which retained its identity while in the act of altering its value. Bolzano points out justly that this 'while in the act' involves a contradiction. It is rather something which embraces a whole 'set of values' all distinct from one another. The fullness of Bolzano's grasp of this point is more evident still from another passage (WL, §69, 314) where he takes special pains to remove misconceptions attendant upon the word 'variable.' We are not to think of an entity undergoing change, but of the possible substitution of one component for another at a given place in a given compound concept or proposition, the candidate for substitution always being chosen from a specified class. In other words, Bolzano is quite clear that in mathematics and in logistics, variability is not changeableness but anyness within a specified set.

2. *Minor disappointments*

1. The number *i* is dubbed an objectless concept 'like zero' (WL, §108) and, though imaginary *qua* concept, does not render imaginary *qua* concept the concept of 'the mathematician who first employed *i*.' And in spite of its contradictions, *i* is and remains a useful idea (WL, §85) and its rejection would cause the loss of beautiful theorems (WL, §447). But for us later folk, beneficiaries of Kronecker, the class ring generated in a polynomial ring by the ideal $[x^2+1]$ stands no more in need of quasi-magical interpretations than does the class ring generated by any other polynomial ideal. Even in Bolzano's day, the other rings isomorph therewith and constructed from points or real number couples by Wessel, Gauss and Hamilton were already to hand, although not already public property.

2. In a passage whose accurate editing is open to doubt (*infra*, page 54) the phrase 'save at isolated points' is inserted as an ornamental precaution, without organic relation to the proof, and three pages afterwards, the same back door is left open in connexion with Taylor's Theorem, which is elsewhere

assumed to be available (DP, xviii) for arcs as functions of abscissae.

3. The sum of an infinite series is described loosely (WL, §84, note) as a number equivalent to the assemblage of all the terms, and moreover as the 'simplest' such number. Even so, Bolzano creditably notes that this would again vary the specification of the term 'sum.'

4. The 'principle of similarity' is too easily supposed applicable (PU, §48) and too easily used for rectifications (DP, 49) and parallels (DD, 208). Doubts thus arise whether the lost memoir 'on parallels' brings any real contribution.

5. Twice in the *Paradoxien* (pages 66 and 100) the *mise en équation* conforms to the Vieta homogeneity canon, and this two hundred years after Descartes.

6. The *Binomischer Lehrsatz* errs (page 47) by underestimating the range of validity of the series at the terminals of the interval of convergence. Ten more years were in fact to elapse before Abel (i) carefully specified the function $(1+x)^{\nu}$, (ii) carefully probed the series with general term $\binom{\nu}{n}x^n$, and (iii) carefully investigated the relations between that function and that series. Only this can truly be called a 'discovery' of the binomial theorem of analysis.

7. Bolzano's apparent betrayals of holomerism have been noted already. It would be idle, and historically insensitive, to hunt for passages where he and Cauchy stumbled for lack of the idea of uniform convergence, or through not perceiving that the real numbers must be preconstructed and not overdrawn on the bank of their future applications.

8. Those who only know Bolzano as a partner in a theorem with Weierstrass, and have not read his earlier and sounder mathematical work, will be particularly disappointed with the footnote to §18, with the psuedo-infinities of §28 and §49, and with the view of complex number stated at the end of §37.

9. Finally, another footnote in the *Paradoxien* appears to belie the great counter-example of Bolzano's own discovery. The point is text-critical as well as mathematical, and will be considered in pages 54–55.

3. *Bolzano's Reception*

In the history of real function theory, Bolzano's counter-example was a feat but not a contribution, since others had become historic when his became known. In fact, the unfortunate circumstances of Bolzano's life after 1820, the denial of further opportunities to inspire academic youth, the enforced retirement far from great libraries, the repeated anonymity and obscurity of publication caused by state censorship, rendered a delay in the recognition of Bolzano inevitable. Even when the *Paradoxien* were rediscovered, they enjoyed too exclusively a *succès de citation*, and two writers even attributed to that work a standard proposition which really belongs to the *Rein analytischer Beweis* (26).

By way of contrast, the logical masterpiece was noticed remarkably soon, though not for its best passages, namely by the Edinburgh Professor of Logic and Metaphysics, Sir William Hamilton (27). On the other hand, a work on 'riddles in mathematics' published in 1945 spoke in Chapter VII almost as if Bolzano had been deceived by the hardihoods of Euler with divergent series. The contemporary inability more clearly to define convergence cannot have been shared by one who constructed the pre-Weierstrass counter-example, anticipated the two-term convergence criterion and died nineteen years later than Abel.

Just so long after his own death, Bolzano was rediscovered by his fellow-mathematicians. In 1871 Hankel (B47) begins to accord him priority in the march towards rigour for series and to praise both his 'unsurpassed acuity' and his 'incisive criticism of earlier work.' In 1872 H. A. Schwarz (B56) refers in passing to his iterated dimidiation process. At length, in 1881, we received a first full-length study, as close as was then possible: from O. Stolz (B57) the rediscoverer of Archimedean-Eudoxian order. The recent discoveries by Jašek take us far beyond the appreciation by Stolz.

But the handsomest tribute to Bolzano came in 1883 from the most competent of pens, from Cantor himself, the inceptor of modern set theory. Cantor calls him '*an extremely acute philosopher and mathematician who set forth his views in the splendid and thoughtful work "Paradoxien des Unendlichen"'* (B46, 560) and this despite some frank criticisms in detail, such as that of Bolzano's definition of a continuum and of his failure to win through to the idea of the cardinal number of a set. Both, says Cantor, are present germinally or in particular manifestations, but fail to emerge as polished concepts. This, Cantor continues, is the explanation of the many '*inconsequentialities and even mistakes in this valuable work.*' The most serious inconsequentiality is, as we saw and shall see, a matter more of textual criticism than of mathematics.

Excellent encyclopedia articles on Bolzano and his mathematics in particular will be found in: *Svensk Uppslagsbok*, 4 (1930), 501, and in *Enciclopedia Italiana*, 7 (1930) 363–364. The Swedish encyclopedia has already become aware of Bolzano's continuous undifferentiable function, whereas the *Britannica* ignores his mathematics altogether.

NOTES TO III: BOLZANO THE MATHEMATICIAN

(1) For 2+2=4 likewise, see Leibniz, with whom Bolzano has so much else in common: *Nouveaus essais*, Book IV, Chapter VII, §10.

(2) By a curious irony, Bolzano himself was the first to construct a continuous curve which is unrectifiable.

(3) Here again, Gauss was the eremitical pioneer in perceiving the problem and going out to solve it: *Allgemeine Lehrsätze usw.*, §§29 ff.

(4) Padova, 3 (1744), 20–22; Albèri, 13 (1855), 35–37; Nazionale, 8 (1898), 77–80.

(5) *De communibus notitiis*, §38=*Moralia* 1079 A. The gibe at the Stoics is a parallel to Bolzano's at Hegel (PU, §12).

(6) *In Euclidis Elementorum Primum*, Leipzig 1873, p. 158: 2–20.

(7) *De aeternitate mundi contra Proclum*, ed. H. Rabe, Leipzig 1899, p. 11: 6–14.

(8) As in (4), but pp. 18–22, 30–37, 73–80 respectively.

(9) Crelle 84 (1878), 242; Acta 2 (1883), 311.

(10) Transactions of the Cambridge Philosophical Society, 11 (1871), 172, signed 2 April 1864.

(11) Cambridge 1903; reprinted 1938, pp. 70, 201, 307, 357.

(12) London 1920, pp. 138–139.

(13) *Beiträge zu einer phänomenologischen Begründung der Geometrie und ihrer physikalischen Anwendungen.* Jahrbuch für Philosophie und phänomenologische Forschung, 6 (1923), 385–560.

(14) *Demonstratio nova theorematis omnem functionem algebraicam rationalem integram unius variabilis in factores reales primi vel secundi ordinis resolvi posse.* Helmstädt, 1799, Werke 3 (1876), 1.

(15) Formerly and puzzlingly called the fundamental theorem of 'algebra,' a name only possible as long as algebra meant everything not elsewhere. In analysis, x enters into rational integer functions (and others) as a variable receiving values, but in algebra, x enters into polynomials as an indeterminate marshalling coefficients. The theorem genuinely corresponding in algebra concerns the existence and interrelation of fields in which complete or partial factorisation takes place. Had the distinction become customary sooner, we might have been spared the anomaly of a fundamental theorem of algebra whose only proofs are analytical.

(16) As in (6) but p. 234: 18–19.

(17) *Journal de l'École polytechnique*, 13 (1806), 154.

(18) *Mémoires de l'Académie de Belgique*, 23 (1872).

(19) *Gesammelte mathematische Abhandlungen*, 2 (1890), 269.

(20) By P. du Bois-Reymond, Crelle, 79 (1875), 29.

(21) G. Darboux records Hermite's distaste for continuous undifferentiable functions.

(22) See A. Pringsheim in *Münchener Berichte*, mathematisch-physikalische Klasse, 27 (1897), 304.

(23) Diels, *Fragmente der Vorsokratiker*, Archytas, Fragment B 4.

(24) *Entwicklung der Mathematik im neunzehnten Jahrhundert.* Berlin 1926, pp. 56–57.

(25) *Grundlagen der Analysis.* Leipzig 1930, p. 4, Satz 4.

(26) W. H. and G. C. Young, *The Theory of Sets of Points.* Cambridge 1906, footnote to §120, p. 191.

(27) *Lectures on Logic*, Edinburgh 1860. These lectures were commenced in 1837, the year of the *Wissenschaftslehre*, and received only minor modifications till Hamilton's death in 1856. For Bolzano, see I, 338 and 344; II, 174.

Venn's *Symbolic Logic*, London 1881, also once alludes to Bolzano, page 426, but only for a variant on the Eulerian diagrams.

IV. Bolzano the Logician

MATHEMATICS was only a fraction of the wide yet perfectly coherent intellectual life of Bolzano. Nay, he was more even than a mathematician felicitously balanced by a sound humanistic education: he was a priest, a theologian, a counsellor of youth, a philosopher and logician, a physicist, a sociologist, a philologist and a writer on the theory of art and beauty. He could be followed into these fields neither without disproportion in the whole nor without great injustice to each. An Introduction to the *Paradoxien* must be content to offer a selected modern bibliography, and briefly to notice two subjects which lie close to mathematics in Bolzano's circle of ideas, Logic and Physics.

1. *Bolzano and Traditional Logic*

The author of the *Paradoxien* is a severe critic of contemporary philosophical literature. As the prime sources of error in philosophy he listed: inadequate conceptual analysis, darkening passions, fluctuation in terminology (BS, 25–36), surreptitious transitions from probability to certainty and the attribution of objectivity to inobjective concepts (1). Compare Whitehead's 'fallacy of misplaced concreteness.' The philosophers of his time, he laments, are loth to define terms; loth to indicate which meaning of an ambiguous term shall hold at which moment; loth to say outright whether a proffered demonstration shall stand alone or be one of several; and loth to make clear which propositions are to be consequents and which are to be antecedents (*Religionswissenschaft*, I, §63, 166–167). In the classical method of mathematics such doubts are not permitted, and Kant did a disservice in suggesting that they could not serve as an ideal even outside mathematics (*ibidem*). Indeed, so great is the degeneration of the time that in mathematics itself we already see writers burking their logical duties!

Bolzano ranged himself among the vigorous opponents of a 'reciprocity between connotation and denotation.' For an account in detail, see Kreibig (B40), and for a clearly chiselled appreciation of Bolzano in this respect, see Scholz (B43, 40).

Bolzano was for long and by many classified as an extreme or as a moderate Kantian. At length, Ueberweg-Oesterreich (2) and others recognised that on certain cardinal points he was a decided adversary of Kant. The entire §79 of the *Wissenschaftslehre* is directed against Kant's doctrine of space and time, and the entire §305 against his doctrine of judgements. Furthermore, Bolzano set up his own distinction between analytic and synthetic judgements, bringing the former (as we shall see) nearly to the status of what we now call tautologies. One of Bolzano's strongest tendencies was, in fact, to alter the very face of logic: more specifically, first to depsychologise it and then to reformalise it.

2. *Bolzano and Modern Logic*

One or two unpretending excursions into algebraical typography would not place Bolzano alongside Leibniz as a pioneer of symbolic or modern or mathematical logic, great as is his affinity with Leibniz at other points. His stronger claim rests more on his deliberated formalisations, and above all on his use in logic of *functions* and *variables* in the strictly mathematical sense.

An inclination to minimise the intrusion of psychology into logic proper is shown in smaller points, such as the vindication of determinateness for infinite sets (§87,* 410–411) on the ground that our being able or unable can never be a criterion objective by nature; but more explicitly in the memoir on *Exner's 'Nominalism and Realism'* (B11). Bolzano congratulates his colleague on taking up the cudgels for an assured content and for exact boundaries to a strictly formal logic, in opposition to contrary tendencies in Hegel, Schleiermacher and others. He is rightly reported by Dubislav (B36) as distinguishing entailments which can be established from intralogical knowledge alone, from such as require extralogical knowledge: as seeking

* In this article all unaccompanied paragraph references are to the *Wissenschaftslehre* of 1837.

a reliable criterion between the two; and as failing for want, on the one hand of an axiomatic, on the other of an algorithmic instrument. The existence of some such wants must have haunted Bolzano's mind when he chid Kant, in the course of his remarks '*On the latest manner of philosophising in Germany*' (*Religionswissenschaft*, I, §63) for calling logic a closed science, and when he subjoined his brilliant prophecy 'that many a fresh discovery remained to be made in logic.'

The logical variable

We saw on page 35 that Bolzano takes 'variability' in mathematics and in logic not literally as *changeableness*, but technically as *anyness* within a specified set. No benign interpretation of a chance phrase, but plain and direct statements of our author afford the authority for this assertion. He is fully aware that numbers of readers will be sure to misunderstand the term 'variable,' but let them note: (i) that in a composite *concept* we are to hold constant all components but one, experiment with substitutions for that one and watch the behaviour of the corresponding composita, say for objectivity and inobjectivity; and (ii) that in a *proposition* we are to hold constant all terms but one, experiment with substitutions for that one and watch the behaviour of the corresponding propositions, say for truth and falsity (§69, 314). In the further course of the *Wissenschaftslehre* there are not a few propositional functions of several variables.

Nor was all this a happy incidental thought and nothing more. It is a deliberate and well-appreciated method. He plans to make frequent use of it (§69, note 2) for a variety of purposes. The width of his programme is suggested by Bolzano's remarks that neither the nature of given propositions (§147, 78) nor that of the relations between propositions (§154, 100) comes fully into view until we vary the arguments and watch the results of that variation.

The difference between a proposition with terms all constant and a proposition with one or more variable terms is brought out in yet another fashion in the paragraph (§147) on the idea of the validity of a proposition. Each individual proposition is either true and then true always, or else false and then false always. Should truth and falsity both be represented, then the reason would always be that some latent variable had not been expressed. Circumstances of place and time were obvious examples. Bolzano here becomes anxious once more lest the term 'variable' be misunderstood, and carefully sends the reader back to the quoted §69 and §108. Such a proposition is 'synthetic' in his sense, not 'analytic' in his sense, namely, not identically true and not identically false.

Sundry particular functions

1. CONJUNCTION. Bolzano's analysis of 'and' as a logical function is recognised, with some modern criticisms, by Russell (*Principles*, 70), but only on the strength of a slighter allusion in the *Paradoxien* (PU, §3) to logical functionality than many others which occur in the fuller treatise on logic.

2. COMPATIBILITY. Bolzano complains (§107) that 'entgegengesetzt' and 'unverträglich' are inadequately distinguished. For compatibility in his sense, two propositional functions must have at least one common argument such that at least one substitution for it renders both true. More is therefore required than that no substitution render one true and the other false.

3. ANTIPODALITY. Two concepts A and B are antipodal (§107) when pure concepts $x, y, \ldots$ exist such that the compositum $(A,x,y, \ldots)$ applies to and only to B, whereas the compositum $(B,x,y, \ldots)$ applies to and only to A.

4. IMPLICATION. Neither our material nor our formal implication is met exactly. Implication is classified as a species of compatibility. It is therefore not enough for no verifier of the antecedent to lie outside the verifiers of the consequent: the two must have (by a qualification inherited from the generic term 'compatibility') at least one common verifier (§§154–155). This restriction is not now imposed, but Bolzano was sharp-sighted enough to notice that on his definition the implication of $g(x)$ by $f(x)$ does not entail the implication of $-f(x)$ by $-g(x)$. More generally, it can happen (§155) that a whole set of propositional functions implies another whole set relative to certain only among the variables.

5. EQUIVALENCE holds, naturally, if each set implies the other: but with the proviso that when the implication is relative to certain variables only, it must be relative to the same ones in

the reverse direction too (§156). With this delicacy we may compare that which qualifies the definition of a subgroup by expressly requiring that the relation in virtue of which the subset is a group be identical with the relation in virtue of which the whole set is a group.

6. ENTAILMENT (Abfolge) is distinguished from implication (Ableitbarkeit) by restriction to 'Wahrheiten an sich,' whereas implication can hold between other kinds of proposition.

7. TAUTOLOGY. Particularly instructive, in the opinion of Bolzano, is the following possible situation: when we vary x through its proposed range we may find that $f(x,a,b, \ldots)$ proves perhaps always true, or perhaps always false, or perhaps always meaningless (§147, 78–79). The next paragraph defines $f(x,a,b, \ldots)$ as analytic if every significance-securing substitution is also a truth-securing one, or again: if every significance-securing substitution is also a falsity-securing one.

8. Bolzano's views on PROBABILITY are today of renewed interest. His discussion in §161 is programmatic by title, since it ranks probability as a *relation between propositions.* It is a kind of relative validity of one proposition with respect to another. As a simple ratio of frequencies or enumerals of favourable cases compared with possible ones, it is rejected.

3. *Language and Symbolism*

Long before logical syntax was opened up, willing acknowledgement was given to the need for correlating grammar and logic. Bolzano did more than make acknowledgements. His three addresses on 'Investigations in the field of the general theory of language,' read to the Bohemian Society of the Sciences in 1844–45, were not printed at the time. Whatever they contained, we are brought near to a genuine philosophical grammar when we turn to Part II, Division II, Section 5 of the *Wissenschaftslehre.*

Its §§169–184 deal with 'some other propositions which on account of their linguistic expression require a special discussion.' They undertake to tease out the more precise logical status of sundry phrases, such as: §170, nothing possesses the quality *b*; §171, a certain *A* possesses *b*; §172, impersonal 'it' and 'one'; §173, a few *A* are *B*, many *A* are *B*; §174, just *n* of the *A* are *B*; §175, *A* possesses *b* in a slighter, in an equal, in a higher degree than *B*; §176, *A* alone is *B*, *A* is *B* alone; §177, *A* is, because *B* is; §178, *A* in its capacity of *C* is a *B*; §179, 'if' and 'then'; §180, *A* determines *B*; §181, 'either' and 'or'; §182, necessary and possible and accidental; and so forth. In the essay on *Exner's Nominalism and Realism* (B11) Bolzano refers the reader back to the *Wissenschaftslehre* for a discussion of 'is' and 'there exists.' Those who wish to explore this extensive logical monument for such matters will be helped by Bolzano's own good thematic index in Volume IV.

In Part V, an elaborate treatise on treatise composition, the grammar and pedagogy of symbolism are discussed with a detail remarkable for the time.

4. '*Sätze an sich*'

Bolzano's epistemology must be remitted to other occasions and other studies of his work (B58 to 63); but since the *Paradoxien* make use of the term, it is desirable at least to set down his own definition of it.

That definition must first be transcribed from the original, since every single-word translation of the operative word would more or less refract the meaning. The formulation in §19 runs in the original thus: 'Unter einem Satze an sich verstehe ich nur irgend eine Aussage, dass etwas ist oder nicht ist; gleichviel ob diese Aussage wahr oder falsch ist; ob sie von irgend jemand in Worte gefasst oder nicht gefasst, ja auch im Geiste nur gedacht oder nicht gedacht worden ist'—'By a "Satz an sich" I mean any statement whatever to the effect that something is or is not, irrespective of whether the statement be true or false, irrespective of whether any person ever formulated it in words, and even irrespective of whether it ever entered into any mind as a thought.'

The depsychologising intention is clear. Pending a better translation, and abstracting from all philosophical criticism, it is therefore suggested that 'Satz an sich' be rendered 'absolute proposition,' to mark the desire of Bolzano for the minimisation if not the omission of all reference to psychological origins and circumstances.

If an absolute proposition is true in fact, Bolzano proposes to call it a 'Wahrheit an sich' (§25) or correspondingly translated: an 'absolute truth.' The component representations (Vorstellungen) of an absolute proposition are in turn to be called 'absolute representations' (Vorstellungen an sich).

Faithful to his existential scruples, already seen fruitfully at work in his mathematics, Bolzano suggested proofs that 'Wahrheiten an sich' exist (§31) and one of them received a new lease of interest from 1903 onwards. It ran as follows: 'It is false that all propositions be false, for in that event this very proposition would be false, and consequently at least one proposition would be true.' The full theory of absolute propositions is developed *ex professo* in Part I, Division II, §§121–194, occupying Volume II, pages 3–326. Bolzano is anxious (§27) to claim the forerunnership of St. Thomas Aquinas, and seeks in their different philosophical preoccupations the reason why their terminologies do not agree more closely.

5. *Expository Maxims*

We have already noticed that the Fifth Part and fourth volume of the *Wissenschaftslehre* is an elaborate treatise on treatise composition. As such, it yields interesting criteria for Bolzano's own treatises, interesting contacts with questions of mathematical methodology and style, and interesting standpoints for reviewers.

The German prose style of the *Paradoxien* shares with other works of its author the old-fashioned leisurely closeness of texture. Long and intricate periods abound, but the reader of stamina who does more than float lazily in a swell of words will detect a logical structure of diamantine clarity. This lack of ambition to write 'brilliantly,' this chosen finespun sobriety, this anxious bordering with qualifications, is caused at its best by Bolzano's intense intellectual eros, his dominant desire to illumine and to communicate, to reason through to the utmost, to stimulate powers of discrimination. It is caused at its worst by weaknesses of which Bolzano is aware in all humility. He roundly accuses himself of a straitened effective vocabulary, of tiresome repetitions, of monotony in the choice of connecting particles, of sentences unduly long and stiff: see the Preface to the *Erbauungsreden* of 1839. He would be grateful indeed for help and advice towards the improvement of his style, though the commonest advice—to read good Greek and German and Latin authors—has hitherto availed him but little.

Turning next to internal canons of exposition, we are not surprised to see the author of the *Wissenschaftslehre* frame a set of rules for the conveyance of truly scientific knowledge. They occur in a manuscript kept in the Prague section of his literary remains, and are printed out by E. Winter (B64, 17–18). Besides revealing Bolzano's ideals, these criteria help us the more judicially to assess the work of recent expositors and the more clearly to see some of the differences between the pre-Bolzano and the post-Bolzano literature of mathematics. Not all his rules can be transcribed in a small compass.

First Rule: In solitary meditation no less than in exposition to others, a clear understanding must be reached as to the meaning of each word and symbol. *Second Rule*: test your concepts for reducibility and irreducibility, and if they are reducible, do not halt the analysis until irreducibile components are found; this is good practice, it discloses valuable facts and it prepares the way for organic rather than adventitious explanations. *Third Rule*: the compounding of concepts from irreducibles must not be arbitrary, and proposed composita must be examined searchingly for consistency and significance. *Ninth Rule*: for scientific knowledge in the narrowest sense, it is needful to show of each fresh assertion either that it is primitive or that it is derivative from stated other assertions, and also to effect the derivation. *Tenth Rule*: an assertion cannot be accepted as a fundamental truth until its truth is originally or demonstratively evident, and until we can show that subject, copula and predicate all are irreducible.

§§512–537 attempt an architectonic of proof as marshalled in a well-composed treatise. One interesting point is, that without demanding it peremptorily, Bolzano strongly desires the *via inventionis* to accompany the *via demonstrationis*.

6. *Appreciations of Bolzano the Logician*

Apart from references by Hamilton and Venn to matters of smaller moment, the logicians took longer to discover Bolzano than did the mathematicians. Erdmann's *History of Philosophy* was for a long time the only work of its kind to notice him to any extent. What Stolz did for him as a mathematician, was done for him as a logician by Husserl.

'It is impossible,' as Palágyi (3) had said a little earlier, 'to convey any idea of the logical subtleties which spring incessantly from the mathematical and combinatorial genius of Bolzano. We can only say that the several thousand pages of the *Wissenschaftslehre* form an unbroken series of unprecedentedly fine distinctions whose full savour is something for which only the palate of a logical gourmet would yearn.'

Husserl (4) says that this treatise is 'a work which in the matter of the theory of the elements of logic leaves the whole literature of systematic logic far behind. Bolzano has not expressly treated the delimitation of pure logic in the modern acceptation, but in his first two volumes he has developed it with such purity of scientific method and with such copiousness in original and ascertained and fruitful thought, that for this reason alone he ought to rank as one of the greatest logicians of all time. Philosophers who have grown up with the habits of thought and language of the idealistic schools will look upon this soberly scientific style as shallowness and lack of ideas, or even as clumsiness and pedantry: but logic as a science must needs build on Bolzano and learn from him what it most of all needs to learn: mathematical acuity in distinction, mathematical precision in theory structure. It will then gain a fresh perspective on the mathematicising tendencies in modern logic; for although Bolzano did not know them as they now stand, they fit naturally into his work.'

Lastly, the *Geschichte der Logik* of H. Scholz (B43), with his historian's justice both to classical and to modern logic and his sharp eye for points of transition, brings several critically enthusiastic appraisals of Bolzano. With Frege, he is one of the greatest two formal logicians in the German literature of the nineteenth century (page 4). He is an early but neglected critic of the reciprocity between connotation and denotation (page 40). And to sum up: 'But all this is overshadowed to insignificance by the work of Bolzano, who has constructed in the first three volumes of his *Wissenschaftslehre* so richly instructive an introduction to logic that one can only say: take and read' (pages 44–45). 'Luminous in themselves are the paragraphs which critically survey previous work on each successive point of doctrine, with particular reference to Aristotle and Kant: in these passages we have an inestimable treasure-house of material for a critical history of logic' (page 45). The principal historical excursus begin with §115, §185, §254 and §717, but minor ones are constantly interspersed.

NOTES TO IV: BOLZANO THE LOGICIAN

(1) From R. Zimmermann's philosophical obituary. Wiener Sitzungsberichte, phil.-hist. Klasse 1849, 163–174; see 167–168.

(2) *Grundriss der Geschichte der Philosophie*, Teil IV. Berlin, 1923, §14 on Bolzano, pp. 175–186.

(3) M. Palágyi: *Kant und Bolzano, eine kritische Parallele.* Halle, 1902, p. 70.

(4) E. Husserl: *Logische Untersuchungen.* Band I: Prolegomena zur reinen Logik. Halle, 1913, pp. 225–226.

V. Bolzano the Physicist

VERSATILE as Bolzano was, we cannot present him as an experimental physicist or as a theoretical physicist on a comprehensive scale. Nevertheless, his works do leave gleanings for the historian of natural science. He debates for example whether a law of continuity holds for natural phenomena, and decides (ZK, §19) that its violation, did a violation take place, would elude observation on principle: for at least two measurements are required and must be separated in time; so that room for continuous transition will always be left. Long and remote as is the road to uncertainty principles, it is not nothing to find in 1842 the envisagement of inobservability in principle, and not merely with regard to absolute motion or for want of more refined instruments.

Space

In a special dissertation, Bolzano tries (DD) to prove that physical space is tridimensional. The proof is difficult to analyse or to follow or to accept at times, but one merit stands out clearly. The conventional reduction to the question of how many mutual perpendiculars can issue from a point is rejected. Bolzano prefers to distinguish 'dependent' and 'independent' groups of points, though without the precision of modern theories of linear dependence and independence over a field, and to show that in physical space we can have four 'independent' points but not five. This being proved, the existence of three and no more mutual perpendiculars from a vertex becomes, as Bolzano would have it, a derived theorem. Möbius' *Baryzentrischer Kalkül* had indeed appeared in 1827, but had found a scarcely better diffusion than Bolzano's own works.

Force

Although force is defined in metaphysical terms (ZK, §17), Bolzano touches incidentally on theorems of dynamics and statics. Using his principle of similarity, that similar determinants entail similar determinands, he undertakes to prove the inertial rectilinearity of free movement: for the path of such a movement must have certain properties, the straight line has them, and no other line has them. The proof is quoted from *Drei Probleme* (§12) where the reader would also find (§10) a 'proof' that, $f(t)$ representing the path, $f'(t)$ and $f''(t)$ represent the velocity and acceleration.

Remitting this curious point to the methodologist of physics, we divert our attention to Bolzano's treatment of equilibrium. He considers at first only a set of n forces along a straight line, and rises into space at the end by means of components. As a first step, the statical problem is analysed into four conditions: (i) each force must be uniquely determined by the rest together; (ii) independently of their sequence; (iii) the continuous variation of all forces but one must induce the continuous variation of the residual force; and (iv) if two equilibrial sets can be conjoined, their conjoint set is also equilibrial; and if one can be removed from the other, the remainders are equilibrial in their own right. Armed with these conditions, Bolzano pursues the reduction into geometry and into analysis. We must seek a sequence of functions specified in turn by $f_n(x_1,x_2,x_3, \ldots x_n)=0$ with $n=1,2,3, \ldots$ answering to the above four conditions of determinateness, symmetry, associability and dissociability (1). After long explorations, during which a function only known to be continuous is unfortunately assumed differentiable (ZK, §52), Bolzano satisfies himself, not only of the obvious fact that

$$f(x_1,x_2,x_3, \ldots x_n)=x_1+x_2+x_3+ \ldots +x_n$$

is a solution, but also that no other solution exists.

Defects notwithstanding, the proof contains an interesting form of catenary inference. The case $n=1$ is vacuous, the cases $n=2$ and $n=3$ are proved separately, the subsequent cases $n=k$ being then proved each from two previous ones, namely $n=3$

and $n=k-1$, all by uniformly conducted appeals to associability and dissociability. The process is undauntedly followed into the infinite, and a vanishing sum declared to be the equilibrium condition even for a set of infinitely many forces along the same straight line (ZK, §52, 31).

Wave Propagation

It is well known that Doppler's Principle was published under a misleadingly narrow title (2) and was received at first with doubt and opposition. But one colleague of Doppler's reviewed it in the next number of *Poggendorffs Annalen* (B29 and 30) and pointed out at once that the principle applied much more widely than Doppler had claimed; in fact, that it held in other forms of wave propagation than light. The same reviewer prophesied that the principle would come to be regularly used in determining the radial proper motions of the stars and for other purposes too. The name of that reviewer was: Bernard Bolzano.

NOTES TO V: BOLZANO THE PHYSICIST

(1) At this point, Bolzano speaks more loosely of 'a' function, despite the circumstance that the number of arguments is left indeterminate, and that functions with different numbers of arguments are supposed somehow to coalesce.

(2) Chr. DOPPLER, *Ueber das farbige Licht der Doppelsterne*. Prague 1842.

VI. The *Paradoxien des Unendlichen*

THE five shorter mathematical works of Bolzano published in his lifetime belong to the period of his active professorate, 1805–20. He had not fallen into disfavour at Vienna, and stood in the prime of life. From his retirement in Těchobus, he wrote a number of mathematical papers for the Bohemian Society of the Sciences, whose titles are given in the *Berichte* for the years 1842–46, and *infra* pages 177–178. Two of them have been edited to modern standards, and an inventory of all the mathematical remains will be found in Gotthardt's dissertation on the problem of truth in Bolzano (B61, page 2 onwards in *second* pagination).

The *Paradoxien* are the work of Bolzano's old age. Indeed, the modern mathematical reader who takes the text as it stands will be occasionally disappointed, and may misjudge Bolzano if he has not also read his earlier work. The qualification of taking the text 'as it stands' is not otiose. Our received version, here translated from the 1851 Leipzig edition, is a posthumous one by a friend with whom Bolzano discussed the topics in question for the last few years of his life. The competence of Přihonský as an editor of mathematical matter and the trustworthiness of the received version of the *Paradoxien* have recently been placed in doubt by M. Jašek, the discoverer of the counter-example.

The antecedents of the *Paradoxien* go back at least as far as 9 June 1842, when Bolzano read 'that part of his paper on the march of ideas to be followed in a truly scientific exposition of mathematics which deals with the finite and infinite.' The next traces are three instalments of papers read as follows: on 9 January 1845 about 'a solution of sundry paradoxes occurring in the mathematical sciences'; on 10 December 1846 about 'calculations with infinite numerical expressions'; and on 24 February 1848 about 'the paradoxes occasioned in mathematics by the idea of the infinite.' The gradual crystallisation of the eventual title is manifest. Between the first and second paper, on

3 February 1845, Bolzano wrote to Přihonský that he was busy with sundry paradoxes in mathematics; between the second and third, he wrote twice to Fesl: on 26 February 1848 that he now realised more than ever the importance of the topics of the *Paradoxien* for mathematics and its philosophy, and on 24 June 1848 that the matter for the *Paradoxien* is constantly expanding under his pen. Finally, on 30 November 1848, only eighteen days before his death, Bolzano read a final instalment under the final title of *Paradoxien des Unendlichen.*

Unlike the manuscript of the *Funktionenlehre*, that of the *Paradoxien* was never made ready for the press by the author himself. That task fell to his devoted but none too mathematical friend Přihonský. In fact, Bolzano himself had entertained doubts about the posthumous treatment of his mathematical papers, for he wrote to Fesl as early as 12 June 1842 that he was convinced 'that his mathematical ideas, in their present form, could not be really well edited by any of his friends.' This premonition on Bolzano's part comes to corroborate the suspicions of Jašek.

Those suspicions were aroused by a palpable contradiction between the *Funktionenlehre* and the *Paradoxien.*

The received text of §37, pages 65–66, as distinguished from the footnote, makes no actual assertion as to the universal existence of derivates of continuous functions; Bolzano simply intends to choose such as are differentiable for a certain purpose: 'Ich begehre nichts anderes, als dass.' But a footnote says it can be shown that 'all well defined functions' are bound to be differentiable 'save possibly for a set of arguments which may indeed be infinite, but whose members must be individually isolated.' No proof is attempted. Jašek (B51, 29–32) claims to possess evidence that the footnote is interpolated. He suspects Slivka von Slivitz, another pupil and friend of Bolzano, of being Přihonský's necessary but not sufficient mathematical counsellor, and of having timorously desired to shield Bolzano from the appearance of mathematical heresy which his counterexample may well have borne to contemporaries.

A fresh critical study of the manuscript is indicated, and external obstacles stand in the way for the present. The difficulty does not, however, reside solely in the passage noticed by Jašek. It recurs in §45, page 88. Strictly speaking, again, there is even here no actual assertion that all continuous functions are differentiable, and the question is further complicated by the admitted fact that the posthumous editor found the manuscript in places rather illegible, in spite of his acquaintance with Bolzano's peculiar abbreviation practices. The presumption of guilt on the part of von Slivitz is slightly enhanced, in Jašek's eyes, by a study of his marginal annotations to a copy of the *Funktionenlehre.* With baffling contrariety, it is also slightly alleviated by the fact that the *Zusammensetzung der Kräfte*, published in 1842 by Bolzano himself—and embodying with acknowledgement some suggestions by von Slivitz—proceeds (§52, page 29) to differentiate a function of which only the continuity is known, together with its satisfaction of conditions which are not quoted as if they *established* the otherwise uncertain differentiability. The hypothesis that Bolzano's mathematical discrimination had become dulled between 1830 and 1848 on this point at least is simple and not to be rejected *a priori*; but neither is it to be accepted easily until renewed and exact archivial research compels us to do so.

The *Paradoxien* originally appeared in 1851, were reprinted in fascimile in 1889 and edited afresh in 1921. See *infra*, page 178.

PARADOXES
OF THE INFINITE

by Dr. Bernard Bolzano

EDITED POSTHUMOUSLY
FROM THE AUTHOR'S MANUSCRIPTS
BY DR. FR. PŘIHONSKÝ

Je suis tellement pour l'infini actuel, qu'au lieu d'admettre que la nature l'abhorre, comme l'on dit vulgairement, je tiens qu'elle l'affecte partout, pour mieux marquer les perfections de son auteur. (Leibniz, *Opera omnia studio Ludov. Dutens*, Tom. II part I. p. 243.

LEIPZIG

C. H. RECLAM SEN.
1851

EDITOR'S PREFACE * iii

THE author of this remarkable treatise on the *Paradoxes of the Infinite* began writing it as early as 1847, during his residence in the company of the Editor at the charming country house in Liboch, near Melnik, but was interrupted by tasks of a different nature and did not complete it until the summer months of the following year, the last of his life. The work not only gave evidence that his intellectual powers (despite his advancing age of sixty-seven years and the visible decline of his bodily strength) had lost nothing of their vigour and alertness, but also fur-
nished the learned world with fresh proofs of the uncommon iv
insight he enjoyed into the most abstract depths of mathematics, natural science and metaphysics. Indeed, had *Bolzano* written and bequeathed to us nothing else than this one treatise, it is our firm belief that on its account alone he would have to be numbered among the most distinguished minds of our century. He knows how to solve with admirable ease the most interesting and complicated of the problems which are raised by the idea of the infinite and have at all times engrossed the attention of workers in these aprioristic sciences; and he can disentangle them before the very eyes of a reader with such clearness that anyone not a complete stranger to the subject, even though he may have hitherto grasped but a few of the matters in question, can still follow the author's exposition, and find at least the majority of his propositions easy to understand. The experts on their part (if only they devote some attention to the treatise, and may we not expect this much from every scholar whatever?) are also bound to notice before long how
important are the ideas which Bolzano suggests in this work and v
develops more circumstantially in others, especially in his *Logic* and in his *Athanasia*, and how he aims at nothing less than a complete transformation of all the modes of scientific exposition hitherto adopted.

* By Dr. Přihonský.

EDITOR'S PREFACE

Upon receiving the manuscript of this treatise from the heirs of the author, the Editor undertook to have it printed as soon as possible, accepting this obligation all the more gladly in that it harmonised with his own inmost feelings; for Bolzano was his unforgettable teacher and friend. He would, moreover, have willingly fulfilled it sooner had his path not been barred by serious obstacles which he was only able to overcome in the course of the current year. He found himself at length in a position to correct his long-standing copy from the sometimes rather illegible and occasionally even incorrect manuscript, to facilitate the use of the book by preparing a full analytical table of contents and to select a satisfactory place of publication. His
vi choice falls on Leipzig, one reason being his hope of a wider dissemination of the treatise, and another being his purpose of doing honour to the celebrated City of Books, the pride and ornament of his adopted Fatherland—for he is by birth a Bohemian. He is confident in fact that once the high genius of Bolzano is generally recognised, it will not be Leipzig's least title to fame to have contributed to the appearance in print of these Paradoxes.

Budissin, 10 *July* 1850.

SYNOPSIS OF CONTENTS

The left-hand column is a literal version of the summary drawn up for the posthumous first edition in 1851. Since Přihonský fails to notice many topics whose importance has since been better realised, his summary is inadequate for the modern reader, and does imperfect justice to Bolzano. The right hand column therefore attempts a more modern analysis.

Editor's Analysis

§1. Why the author intends to deal with no other paradoxes than those of the infinite.

§§2–10. The idea of the infinite as conceived by mathematicians, and a discussion of that idea.

Translator's Analysis

§1. Infinity is the source of vii
most if not all mathematical and allied paradoxes, and the first necessity is to define it.

§2. Approaches to a definition: criteria for finite and infinite sets will cast light on the infinite as such.

§3. The logical constant 'and,' together with the notion of an 'aggregate' (*Inbegriff*) as derived from logical conjunction.

§4. 'Sets' (*Mengen*) are aggregates considered as indifferent to permutation. 'Multitudes' (*Vielheiten*) are sets whose members are co-specific, and 'unities' (*Einheiten*) are cospecific entities considered as members of a set.

§5. Sets can be main sets composed of subsets, and are called 'sums' (*Summen*) when subset members can equally well be regarded as mainset members. Associativity enters into the very definition of a sum.

Editor's Analysis	*Translator's Analysis*
§§2–10. The idea of the infinite as conceived by mathematicians and a discussion of that idea.	§6. 'Quantities' (*Grössen*) are members of sets such that each two of their members are either identical, or else stand to one another in the relation of logical sum to logical summand.
	§7. When a fixed one-one relation exists, such that each member of an aggregate is either a prerelatum with a unique fellow-member as postrelatum, or else a postrelatum with a unique fellow-member as prerelatum: then the aggregate is called a 'sequence' (*Reihe*) and its members are called the 'terms' (*Glieder*) of the sequence. The fixed relation is called the 'law of construction' (*Bildungsgesetz*) while the pre- and postrelatum are called the 'antecedent' and 'consequent' (*vorderes und hinteres Glied*). Terms having both an antecedent and a consequent are called 'inner' (*innere Glieder*) and terms, if any, which lack the one or the other are called 'outer' (*äussere Glieder*) being either the 'first' if any or the 'last' if any.
	§8. Let the first term of a sequence be a unity of given species, and let each following term be a multitude formed by adjoining to the antecedent term some yet unadjoined but co-specific unity. Then each such following term is called a 'finite' (*endlich*) or 'countable' (*zählbar*) multitude. The name 'number,' or more definitely 'whole number' (*Zahl, ganze Zahl*), shall apply both to each such following

Editor's Analysis	*Translator's Analysis*
	term and also to the initial unity. Aliter: whole numbers are finite if they can be reached from 1 by terminated mathematical induction.
	§9. Multitudes would be called 'infinite' (*unendlich*) if every finite multitude were equipotent with a proper part of them. Their existence will be proved in the sequel.
	§10. The infinite as such, being infinite only in virtue of some latent infinite multitude, is hereby incidentally determined.
§11. How Hegel and other philosophers conceive the infinite.	§11. Hegel and others rightly reject the infinite as an unbounded variable, but wrongly allow it in cases other than those of latent infinite multitude.
§12. Other definitions of the infinite and the judgement to be passed upon them.	§12. Even mathematicians err in taking the infinite as the limit of unlimited increase, as the incapacity of further increase, as the absence of termination, or as that which exceeds any assignable magnitude.
§13. The objectivity of the concept set up by the author is evinced by examples from the realm of the non-actual. The set of all 'absolute propositions and truths' is infinite.*	§13. The existence of infinite sets is evinced by an example from logic in parallel with the above notion of number.
§14. The refutation of some objections urged against this concept.	§14. Answers to the objections that infinite sets can never be completed wholes, and that aggregates arise solely through our actual or possible mental acts. The logical analysis of impossibility.

* 'Wahrheiten und Sätze an sich,' a characteristic term in Bolzano's philosophy. See here page 45.

Editor's Analysis	*Translator's Analysis*
§15. The set of all natural numbers is infinite.	§15. Answers to the objection that the set of all natural numbers is enumerated by its last member and thus is finite instead of infinite.
§16. The set of all quantities whatever is an infinite set.	§16. The set of all quantities is with still greater reason an infinite set—Bolzano counting our 'positive real numbers' (rational and irrational) as quantities but not as numbers.
§17. The set of all the simple parts constituting either space or time at large, or any closed spatial or temporal interval, is an infinite set.	§17. Further examples of infinites are: the points or instants in space or time as a whole, or in any closed spatial or temporal interval. These examples are non-actual, being neither substances nor qualities.
§18. Not every quantity envisaged by us as the sum of infinitely many other quantities, each of which is finite, is itself an infinite quantity.	§18. The 'sum' of infinitely many individual finite terms need not be infinite. Proposals for a more rigorous summation of convergent geometrical series.
viii §19. There exist infinite sets which are greater or less than other infinite sets.	§19. One infinite set can exceed another as to multitude, namely by being a whole of which the other is a proper part.
§20. A remarkable relation between two infinite sets, consisting in the possibility that each member of the one set can be so coupled with a member of the other that no member in either set remains uncoupled, and no member in either set occurs in two or more of the couples.	§20. Biunivocal correspondence can obtain between the members of an infinite main set and those of an infinite proper subset, with two examples.*
§21. Despite their equinumerosity in members (in Hinsicht auf die Vielheit ihrer Teile	§21. Conversely, infinite sets in holomeric correspondence are not necessarily equal as multi-

* Bolzano coins no technical term for this situation, and repeats its description when required. In the present volume we shall speak briefly of 'holomeric correspondence' and more briefly still of 'holomerism.'

Editor's Analysis

gleich) infinite sets can still be unequal as multitudes (Ungleichheit ihrer Vielheiten) in that the one turns out to be a proper part of the other.

§§22–23. Why the case is different with finite sets, and why the grounds of difference cease with infinite ones.

§24. Even when two quantities are sums of infinite sets of addends equal in corresponding pairs, we are not permitted for that reason alone to equate them, but only when the criteria of membership for the two sets are identical.

§25. Infinites exist in the realm of the actual as well as in that of the non-actual.

§26. This assertion is not contradicted by the principle of the 'universal determinateness of the actual.'*

§27. Nevertheless, those mathematicians err who speak of intervals of time terminated on both sides and yet infinitely great or (more often) infinitely small, or of infinitely great and small distances, and so do physicists and metaphysicians who presuppose or assert that some forces in the universe are infinitely greater or less than other forces.

Translator's Analysis

tudes, but can also stand to one another in the relation of whole and part.

§22. This apparent paradox arises only because holomerism does, in the case of finite sets, entail equimultitude.

§23. Whereas the holomerism of infinite sets, besides forfeiting this entailment, is accompanied by other grounds of dissimilarity.

§24. We may not equate the two 'sums' of two infinite sets of addends on the sole ground that these addends are equal in corresponding pairs, for the two sets must also agree in their criteria of membership.

§25. Actual as well as non-actual infinites exist, with examples from theodicy.

§26. The principle of the universal determinateness of the actual* is not violated by this thesis.

§27. Some infinites falsely assumed by the mathematicians themselves are: forces, and closed intervals of time or space, supposed to be infinitely greater or (more often) infinitely less the one than the other.

* 'Grundsatz der durchgängigen Bestimmtheit alles Wirklichen,' a characteristic maxim in Bolzano's philosophy, with the status of an ontological counterpart to the logical principle of excluded middle.

Editor's Analysis	*Translator's Analysis*
§28. The chief paradoxes of the infinite in the domain of mathematics; first of all, in the general theory of quantity, and especially in the doctrine of number.	§28. Computation with infinites is paradoxical only in appearance, because it does not undertake to determine single infinites by means of finite numbers, but only to take cognisance of special relations between different infinites; and this latter is quite possible.
§29. A computation with the infinitely great really does subsist.	§29. Examples of computation with infinites, and precautions against false analogy with the finite.
§30. And so does a computation with the infinitely small.	§30. Examples of computation with 'infinitesimals.'
ix §§31–32. The falsity of some ideas which even mathematicians entertain of the infinitely great and the infinitely small.	§31. Some errors still current among mathematicians, especially with regard to the pretended quotients 1/0 and 0/0. §32. Some errors in the pretended summations of $\Sigma(-1)^n a$, which is a symbol not expressing any true quantity at all.
§33. The caution to be exercised when computing with the infinite, if we are to avoid going astray.	§33. Further warnings: the addends of one series can form a proper subset of the addends of another whose terms are all positive, and yet the sum of the former infinitely exceed the sum of the latter.
§34. A more precise determination of the idea of zero. Zero may never be used as a divisor in an equation purporting to be more than a mere identity.	§34. Zero cannot be defined as 'representing nothing,' but only by such equations as $A-A=0$, $A+0=A$, $A-0=A$ and $0A=A0=0$. It is to avoid inconsistency with these equations that we must forbid division by 0.
§35. Contradictions arising from the occasionally encountered assertion that infinitely	§35. The unlawfulness of equiparating 'relatively infinitesimal' with 'equal to zero' is

Editor's Analysis

small quantities, upon entering into additive or subtractive relations with certain others, become zero or vanish.

§36. These contradictions are not evaded by the assumption of some mathematicians that infinitely small quantities are mere zeros and infinitely large ones are quotients resulting from the division of an infinite (*sic*) quantity by a mere zero.

§37. How the author conceives himself obliged to regard the method of computation with the infinite in order to free it from all contradiction.

§38. Paradoxes of the infinite in the applied part (*sic*) of the theory of quantity, and specifically in the theory of time and space. Apparent contradictions lurk in the very idea of a continuous extension or a continuum, and how they can be resolved.

§39. Paradoxes in the idea of time.

Translator's Analysis

made plain by the comparison of certain rational with certain irrational cases.

§36. At a common zero of $f(x)$ and $g(x)$, the fraction $f(x)/g(x)$ may have a limit, but cannot have a value; and its pretended evaluations involve a latent and unlawful division by zero.

§37. These errors being duly censured, the author's positive principles are: (i) the variables need not become 'infinitely small'; (ii) the functions must all possess derivates; (iii) with suitable precautions and interpretations the differential notation is allowable; and (iv) save for 'isolated points' in finite or infinite number, all 'well-definable' (*überhaupt bestimmbar*) functions do possess derivates, and admit finite Taylor series with mean value remainders.

§38. Other paradoxes occur in the applications of the theory of quantity, notably with regard to the doctrines of time, space and the continuum. The classical arguments for and against the composition of continua from indivisibles, and Bolzano's decision in favour.

§39. Opinions on the ontological status of time. Instants as prerequisite to the principle of excluded middle, and time as the set of all instants.

Editor's Analysis	*Translator's Analysis*
§40. Paradoxes in the idea of space.	§40. Opinions on the ontological status of space. Points as determinants of kind and amount of change, and space as the set of all points. The volume of a spatial region is not governed by the mere multitude of its points, but by an axiom securing isomorphism between the juxtaposition of volumes and the addition of their measuring numbers.
§41. How the author's idea of space serves to explain most of the contradictions in the doctrine of space.	§41. Seven strange, but now not contradictory propositions about open and closed intervals, pairs of sets of points standing in an irrational ratio, and the proportionality of numerosity and magnitude in the case of geometrically similar sets of points.
§§42–43. How a false interpretation of the doctrine of the infinitely great has led some mathematicians to form incorrect notions.	§42. Other writers have hitherto touched only the last of these points, and then only to deny it by an argument wrongly inferring the equality of two sets from their mere holomerism. §43. Mathematicians frequently violate the principle that each two points in space stand at a finite distance. In the case of tangents and secants, this violation revenges itself by confusion as to signs.
§44. J. Schulz's calculation of the volume of infinite space, and wherein the real error of this calculation lay.	§44. By treating infinite radii as finite, J. Schulz mistakenly computed the volume of the whole of space at $\frac{4}{3}\pi\infty^3$, and similar criticisms expose the futility of the spurious paradoxes advanced by Boscovich.

Editor's Analysis	*Translator's Analysis*
§45. Many an absurd statement has also been occasioned by the doctrine of the infinitely small.	§45. 'Infinitesimals' of the first and higher orders have been assumed scarcely less often, the existence of derivates supplying, especially in geometry, an undeserved protection against erroneous results.
§46. What we are to think of Galilei's proposition that the circumference of a circle is equal in size to its centre.	§46. A discussion of Galilei's deliberately absurd 'proof' that the circumference and the centre of a circle are equal in size.
§47. An elucidation of the theorem that the common cycloid has an infinite curvature at the point where it reaches down to its base line.	§47. Another difficulty with x 'infinitesimals,' concerning the curvature attributed to the common cycloid at a cusp.
§48. How it comes about that many extended figures stretch into infinite space yet remain finite in magnitude, that other figures though spatially confined are infinite in magnitude, and others again remain finite despite an infinity of whorls round a fixed pole.	§48. Some anomalies in the finitude or infinitude of geometrical entities compared with their dispersal or confinement in space.
§49. Some other paradoxical features of spatial extensions infinite in magnitude.	§49. Nineteen miscellaneous cases of the comparison or enumeration of sets of points.
§50. Paradoxes of the infinite in the realm of physics and metaphysics. What truths must be recognised in order rightly to judge these paradoxes. A proof that no two entities can be exactly similar, and hence that no two exactly similar atoms (simple substances) can exist in the universe.	§50. Turning to physics and metaphysics, Bolzano lays down: (i) the non-existence of any two completely similar entities, (ii) the necessary existence and variability of simple substances and (iii) the insufficiency for physics of experience taken alone and not eked out by conceptual truths. Opposition from the physicists is only to be expected.
§51. Some prejudices to be dismissed before the relevant paradoxes can be correctly weighed up. There is no such	§51. For certain prejudices must be cleared away. Firstly: the prejudice that there exist purely inert matter. The lawfulness, but

Editor's Analysis	*Translator's Analysis*
thing as dead and purely inert matter.	also the dangers, of simplificatory hypotheses.
§52. It is a prejudice of the schools that the hypothesis of an immediate action of one substance on another is outruled.	§52. Secondly: although the immediate action of one substance on another may not be asserted without proof, neither may it be denied *a priori*. How this led Leibniz to his pre-established harmony.
§53. It is likewise a prejudice to believe that an immediate action at a distance is impossible.	§53. Thirdly: we must not only reject the denial of *actio in distans*, but turn right round and maintain that every action of one limited substance on another is an *actio in distans*.
§54. The interpenetration of substances must be categorically denied.	§54. Fourthly: the interpenetration of substances, sometimes asked for in chemistry, must be categorically denied, because it would make physical states incalculable.
§55. The prejudice that spiritual beings are totally unspatial, to the extent of being unable to occupy the place of a single point. The only differences between created substances are differences of degree.	§55. Descartes' dualism separates space and spirit too drastically, and Kant's doctrine of forms of intuition intensifies the evil.
§56. The great paradox of the bond between spiritual and material substance is automatically resolved when this view is taken.	§56. Bolzano's own theory solves the problem of how spirit and matter can interact, though it must be confessed that much of this still awaits discovery.
xi §57. The erroneous idea of a construction of the universe from forces alone and without substances.	§57. Pure dynamism must be rejected, and we must admit the existence of one uncreated and infinitely many created substances.
§58. In all God's creation there is neither a highest nor a lowest degree of being.	§58. God's creation is infinitely graded, and each single grade is at each single moment occupied by some creature or creatures.

Editor's Analysis

§59. Differences in the density of bodies are quite compatible with the continuous filling of infinite space with substances, and it is not necessary to assume an interpenetration of substances.

§60. Every two substances in the world are continually in a state of interaction.

§61. Among them there exist 'dominant' substances, but none of them is endowed with forces infinitely greater than those of the 'dominated' ones.

§62. Whether every aggregate of substances require to contain one and* only one member of maximum strength.

§63. Over and above the dominant substances there exists another world-stuff, the ether, which fills the space between them, and binds all bodies together without itself containing dominant substances. Attraction and repulsion take place between substances, and what notion the author entertains of them. How it comes about that kinds of matter which differ as to force, and above all as to the force of their mutual attraction, are for all that universally equal in weight—in other words, that their weights are proportional to their masses.

§64. How the dominance of certain substances or atoms over

Translator's Analysis

§59. In virtue of holomerism, the occupation of each point in space by just one corresponding substance is quite compatible with an infinite gradation of density.

§60. Every two substances continually interact, and the change induced in any one by all the rest in concert is a continuous change.

§61. Yet action can exceed reaction, giving rise to 'dominant' and 'dominated' substances. Their difference in force, however, is never infinite.

§62. Though finite sets of substances must contain members of maximum strength, infinite sets are under no such necessity.

§63. Each finite region of space contains but finitely many dominant substances. Bolzano's theories of physical bodies and intervening ether, of attraction and repulsion, of a threshold-restricted law of gravitation, of the proportionality of weight to mass, and of a caloric substance both ponderable and also weight-conferring.

§64. The theory of dominant substances explains a certain

* Dr. Přihonský prints 'Eine,' not 'eine.'

	Editor's Analysis	*Translator's Analysis*
	others is manifested, and what its consequences are.	apparent mutual repulsion, for whose explanation no recourse is necessary to a primitive force of repulsion between ether atoms.
	§65. No dominant * substance suffers so great an alteration as to be set free from every part of its immediate surroundings.	§65. No dominant substance is ever so completely changed as to lose the whole of its envelope of dominated substances.
	§66. Where one body ends and another begins, that is, the problem of the boundaries of bodies.	§66. Considerations from this point of view on the nature of the physical frontiers of physical bodies.
	§67. Whether, and when, bodies stand in immediate contact with one another.	§67. Corresponding considerations on the nature of the physical contact of physical bodies.
	§68. The possible kinds of movement taking place in the universe.	§68. By reconciling the plenum with variable density, the author's theory permits many other kinds of movement than that in which constituent particles simply interchange positions—in particular, oscillations and rotations. For a reason too obvious to be given (*sic*) all ether atoms and almost all dominant ones must continually oscillate.
xii	§69. Whether any atom in the universe ever describes a perfectly straight or a perfectly (*sic*) curved line. Do the author's views on the infinitude of the universe leave room for any definite translational or rotational movement of the universe as a whole?	§69. Perfectly circular or rectilinear movement is indeed possible with atoms, but shall not be asserted ever actually to happen. Concerning movement in a straight-sided elbow and in a logarithmic spiral. Is the infinitude of the universe compatible with any movement of the universe as a whole? One contrary reason is invalid, another is valid.

* The context shows that Dr. Přihonský's 'ausgezeichnete Substanz' is not a new term, but merely a literary variant on 'herrschende Substanz.'

Editor's Analysis

§70. Two paradoxes made famous by Euler.

Translator's Analysis

§70. Two final paradoxes: (i) in planetary theory, the cases of simple release and of projection toward the centre of force are discrepantly solved by Euler and by Busse, but Bolzano dismisses the discrepancy as based on a physically false hypothesis; and (ii) an apparent anomaly between the times of descent down a pendular arc and down the corresponding chord.

PARADOXES OF THE INFINITE

§1

CERTAINLY most of the paradoxical statements encountered in the mathematical domain—though not, as *Kästner* (1) would have it, all of them—are propositions which either immediately contain the idea of the *infinite*, or at least in some way or other depend upon that idea for their attempted proof. Still less is it open to dispute that this category of mathematical paradoxes includes precisely those which merit our closest scrutiny, inasmuch as a satisfactory refutation of their apparent contradictions is requisite for the solution of very important problems in such other sciences as physics and metaphysics. 1

This, then, is the reason why I address myself in the present treatise exclusively to the consideration of the paradoxes of the infinite. As is readily understood, however, it would be impossible to recognise the appearance of contradiction in these paradoxes for what it is, namely a mere appearance, unless we first of all became quite clear what precise notion we attached to the term infinite. Consequently, this point shall be taken first.

§2

The very word infinite shows that we put the *infinite* into contrast with the merely *finite*. Again, the derivation of the former *name* from the latter betrays the additional fact that we consider the *idea* of the infinite to arise from the idea of the finite by, and only (2) by, the adjunction of a new element; for such in fact is the abstract idea of *negation*. And finally: it is already one reason against denying the application of both ideas to *sets* (more precisely, to *multitudes*, that is, sets of unities) and hence to *quantities* as well, that *mathematics*, the doctrine of quantity, be the very place where we most often speak of the 2 infinite; for we there select as matter for consideration and even for computation both finite and infinite multitudes, as also both *finite* and infinite magnitudes, the latter including not only the *infinitely great* but also the *infinitely small.* Quite independently of the hypothesis that these two ideas, of the finite and of the infinite, are applicable only to objects which in some respect or other exhibit *quantuplicity* or *quotuplicity*, we are already entitled to hope that a rigorous investigation of the circumstances in which we pronounce a set to be finite or to be infinite will also afford us information about the nature of the *infinite as such.*

§3

For this purpose, however, we must go back to one of the simplest conceptions in our minds and seek agreement on the meaning of the word by which we propose to designate it. It is
3 the conception underlying the conjunction *And*, whose most appropriate expression, enabling it to come to the forefront as clearly as the aims both of philosophy and of mathematics demand in numberless cases, is found to my best belief in the words: 'An aggregate of well-defined objects' (ein Inbegriff gewisser Dinge) or: 'A whole composed of well-defined members.' (3) We intend these words to be interpreted so widely that, whenever the conjunction *And* is customarily employed, for example in the sentences:

'*The sun, the earth and the moon act upon one another*,'

'*The rose, and the conception of a rose, are a couple of entirely distinct entities*,'

'The names '*Socrates*' and '*son of Sophroniscus*' designate one and the same person,'

we can say that the subject of these sentences is an *aggregate of well-defined objects*, or *a whole composed of well-defined members*; and say in the first example, that it is the whole composed of the sun, the earth and the moon as members which we pronounce to be a whole whose members act upon one another; in the second example, that it is the aggregate composed of the object 'the rose' and the object 'the conception of a rose' as members which we pronounce to be an aggregate whose members are entirely distinct entities; and so forth. These few remarks will presumably be quite sufficient for a common understanding of the conception here dealt with, at least if we add this statement: that every arbitrary object *A* whatever can be united with any other arbitrary objects *B*, *C*, *D* . . . whatever to form an aggregate, or to speak still more rigorously, that these objects already form an aggregate without our intervention (an sich selbst schon) and one about which numerous truths of different degrees of importance can be enunciated—provided only that *A*, *B*, *C*, *D* . . . really represent, each and every one of them, a *distinct* object; that is, provided only that none of the propositions '*A* is identical with *B*,' '*A* is identical with *C*,' '*B* is identical with *C*' and so forth be true. For if *A* were identical with *B*, it would of course be absurd to speak of an aggregate composed of the objects *A* and *B*.

§4

There exist aggregates which agree in containing the selfsame 4
members, and nevertheless present themselves as *different* when seen under different aspects (Gesichtspunkte) or under different conceptions (Begriffe), and this kind of difference we call 'essential' (wesentlich). For example: an unbroken tumbler and a tumbler broken in pieces, considered as a drinking vessel. We call the ground of distinction between two such aggregates their *mode of combination* (Art der Verbindung) or their *arrangement* (Anordnung). An aggregate whose basic conception renders the arrangement of its members a matter of indifference, and whose permutation therefore produces no essential change from the current point of view, I shall call a *set* (Menge), and a set whose members are considered as *individuals* (Einheiten) of a stated species *A* (that is, as objects subsumable under the concept *A*) is called a *multitude* (Vielheit) of *A*.

§5

Some aggregates, it is well known, have members which are themselves composite, themselves aggregates in turn. Among these again, some are regarded from a viewpoint which renders it indifferent whether or no we take the members of the sub-aggregates as members of the main aggregate (die Teile der

Teile als Teile des Ganzen selbst auffassen). I borrow a term from mathematics, and call such aggregates *sums*. For it is the essential nature of a sum that $A+(B+C)=A+B+C$ (4).

§6

If we regard an object as belonging to a category of entities such that each two of them, say M and N, can stand in no other mutual relation than that either they are *equal*, or else one of them can be represented as a sum whose augend is equal to the
5 other of them (in other words, either $M=N$, or else $M=N+\nu$ or $N=M+\mu$, where ν and μ must again be either equal or such that one can be considered as a part contained in the other), then we regard that object as a *quantity*.

§7

The name *series* (Reihe) shall be given to a proposed aggregate of objects . . . $A,B,C,D,E,F,$ . . . , $L,M,N,$ . . . when it is possible to assign to any one member M exactly one other member N by a *law applying uniformly* to all members and *determining*, either N through its relation to M, or M through its relation to N. The members shall be called the *terms* (Glieder). The law whereby M determines N or conversely shall be called the *principle of construction* (Bildungsgesetz) of the series. One of the two terms (whichever you like, but without wishing to imply any temporal or spatial sequence) shall be called the *antecedent* (vorderes Glied) and the other the *consequent* (hinteres Glied). Every term M which has both an antecedent and a consequent (and which therefore both takes its rise by the principle of construction from another member, and in turn gives rise to a third) shall be called an *inner term* (inneres Glied) of the series. Hence, finally, it will be plain which terms (if any of the kind exist) shall be called *outer*, which *first*, and which *last*.*

§8

6 Let us imagine a *series* whose *first* term is an *individual* of the species A, and whose every subsequent term is derived from its predecessor by adjoining a fresh individual of the species A to

* More detailed explanations of these ideas, and of some others which have occurred in previous paragraphs, will be found in the *Wissenschaftslehre*.

form a sum with that predecessor or its equal. Then clearly will all the terms of this series, with the exception of the first, which was a *mere individual* of the species *A*, be *multitudes of the species A*. Such multitudes I call *finite* (endlich) or *countable* (zählbar) *multitudes*, or quite boldly: *numbers*; and more specifically: *whole* numbers—under which the first term shall also be comprised.

§9

When we vary the nature of the concept here denoted by *A*, we shall find that the objects subsumed under it, the individuals of the species *A*, form now a more numerous and now a less numerous set; and the series under consideration will contain now a more numerous and now a less numerous set of terms. In particular, the series may contain so many terms that it cannot, compatibly with taking in and exhausting *all* the individuals of that species, be conceded to have a *last term*; a point which we shall handle in greater detail in the sequel. Assuming it for the present, I propose the name *infinite multitude* for one so constituted that every single finite multitude represents only a part of it.

§10

I hope it will be conceded that this definition of a *finite* and an *infinite* multitude will distinguish between them exactly in the 7
sense intended by those who have used these terms rigorously. It will also be granted that no vicious circle lurks in the definitions. It only remains to ask, therefore, whether a definition merely of what is to be called an infinite *multitude* can put us in a position to determine what the *infinite* be in itself. That would be the case if it came to light that multitudes were the only things to which the idea of the infinite could be applied in its strict sense —in other words, if it came to light that infinitude were, strictly speaking, a property only of multitudes; or again in other words, if everything we judge *infinite* is so judged solely because, and solely in so far as, we find it possessing a property which can be considered as infinite multitude. Now to my thinking, this really is the case. Mathematicians use the word plainly in no other meaning: for what they are occupied in determining is hardly ever anything else than quantity, and they do so by first choosing one object of the species as *unit*, and then employing

the idea of number. If they find a quantity greater than any finite number of the assumed units, they call it *infinitely great*; if they find one so small that its every finite multiple is smaller than the unit, they call it *infinitely small*; nor do they recognise any other kind of infinitude than these two, together with the quantities derived from them as being infinite to a higher order of greatness or smallness, and thus based after all on the same idea.

§11

8 Some philosophers, however, and notably in our day, such as *Hegel* and his followers, are not satisfied with this mathematically familiar infinitude. They contemptuously dub it the 'bad infinity,' and claim knowledge of a true, a vastly superior, a *qualitative infinity*, to be found in *God* particularly, and speaking generally only in the *absolute*. Now I agree for my part with *Hegel*, *Erdmann* and others so long as they are thinking of the mathematically infinite only as a *variable* quantity knowing no limit to its growth (a definition adopted, as we shall soon see, even by many mathematicians) and so long as they are finding fault with this notion of a quantity which is always *growing* into the infinite but never *reaching* it. In fact, a *truly infinite* quantity (for example, the length of a straight line unbounded in either direction, meaning: the magnitude of the spatial entity containing all the points determined solely by their abstractly conceivable relation to two fixed points) does not by any means need to be variable, and in the adduced example it is in fact not variable. Conversely, it is quite possible for a quantity merely capable of being taken greater than we have already taken it, and of becoming larger than any one pre-assigned (finite) quantity, nevertheless to remain at all times merely finite: which holds in particular of every numerical quantity 1, 2, 3, 4, . . .; what I refuse to admit is only this: that the philosopher knows any object to which he is entitled to attach the predicate of infinitude without having first established that in some respect or other that object exhibits infinite quantity, or at least infinite multitude. Now if I can once show that in God Himself, the Being Whom we regard as the most perfectly one, aspects can be found under which we see even in Him an infinite multitude, and if I can show that we attribute infinitude to Him under

those aspects alone, then it will scarcely be necessary to go on and show that similar considerations lie at the bottom of all the other cases where the idea of the infinite holds good. I say then: 9
we call God infinite because we are compelled to admit in Him more than one kind of force possessing infinite magnitude. Thus, we must attribute to Him a power of knowledge which is true omniscience, and which therefore comprises an infinite set of truths, to wit, all truths—and so forth. And what idea of the truly infinite would people impose upon us other than the one here set up? They say it is that All which comprises every Something whatever, the absolute All outside of which there exists nothing more. Yet even on this formulation it would still be an infinite which according to our definition included an infinite multitude. It would be an aggregate not alone of all actual entities, but also of all things having no actuality, of all 'absolute propositions and truths.' Consequently, there appear to be no grounds for departing from our conception of the infinite and embracing theirs—to say nothing of all the other errors they have woven into this doctrine of the All.

§12

On the other hand, I am obliged to reject as erroneous many other definitions of the infinite, even such as have been set up by mathematicians in the belief that they were doing nothing but exhibit the elements of this one same concept.

1. As I mentioned only a short while ago, some mathematicians (including even *Cauchy* in his 'Cours d'analyse' and several other works, and the author of the article 'Infinite' in *Klügel's* Dictionary) imagined they were defining the infinite when they described it as a variable quantity whose value increases *without limit* and in keeping therewith is capable of surpassing *any one pre-selected quantity* (gegebene Grösse) *no matter how large* a one has been chosen (noch so gross). The *limit* of this unlimited growth is alleged to be the *infinitely great quantity*. In this manner the tangent of a right angle, regarded as a continuous quantity, is supposed to be unlimited, unterminated, 10
infinite in the strict sense. How mistaken this opinion is can be seen from the simple fact that what mathematicians call a *variable quantity* is not properly a quantity at all, but only the idea and notion of a quantity, and a notion, moreover, which comprises

under itself not a fixed individual quantity at all, but rather an infinite set of quantities, distinct from one another, and differing from one another in their values—which is precisely the same thing as saying that they differ from one another in *quantity*. The thing they call infinite is not, however, any one of the *distinct* values taken on in their example by *tan* (φ) for distinct values of φ, but rather the single value which they imagine (wrongly as it here happens) *tan* (φ) to take on for $\varphi = \frac{1}{2}\pi$. Then again, it is surely another contradiction to speak of the limit of an unlimited growth, and when defining the infinitely small to speak of the limit of an unlimited diminution; and if the former is declared to be the infinitely great, then analogy demands that the latter, namely zero pure and simple (just Nothing) be declared to be the infinitely small: which is surely incorrect, and which neither *Cauchy* nor *Grunert* ventures to assert.

2. The definition just considered was too wide. Another definition is too narrow, namely the one adopted by Spinoza and many other philosophers, as well as mathematicians, according to which *those things alone are infinite which are incapable of further increase*, or to which nothing can be adjoined or added. Mathematicians hold themselves entitled, in fact, to add other quantities to any quantity whatever, even to one infinitely great, and not only to add finite ones, but also to add other quantities which are already themselves infinite, and even to multiply one infinity by another infinity, and so forth. And granting that the legitimacy of this process is disputed by some,
11 what mathematician is there who, if he allows infinity of any kind, is not forced to concede that the length of a straight line bounded on one side but stretching to infinity on the other is infinitely great and nevertheless capable of being increased on the side hitherto limited?

3. No whit more satisfactory is the definition given by those who lean on the derivation of the word and say: the infinite is that *which has no end*. If they are thinking in this definition only of an end in time, only of a cessation, then no other things could be infinite but those subject to temporal flux, whereas we also ask of things not so subject, like lines or abstract quantities, whether they be finite or infinite. But if they take the word end in a wider sense, say as equivalent with *limit* as such, then I call attention to two points:

Firstly, there are many objects which, while we cannot well prove them to possess a limit without foisting on that word a highly vacillating and thoroughly confusing significance, are nevertheless classified by nobody whatever as infinite. For example, a simple point of time or space *has* no such thing as a limit, but rather is itself the limit of an interval of time or of line, most people defining it precisely as such, just as if this were its very essence; but for all that, it never occurred to anybody (with the possible exception of Hegel) to espy infinitude in a mere point. Or for another example: mathematicians know of no bounding point in the circular periphery, of no bounding point or line in the numerous other closed curves and surfaces, and consider them for all that to be finite objects—unless indeed they come to speak of the infinite set of points contained in them, from which point of view, however, the like infinitude must be recognised in every bounded rectilinear segment.

Secondly, I observe that plenty of objects are undeniably bounded and yet are still regarded as belonging to the class of 12
infinite quantities. Such is the case not only with the unilaterally unbounded straight line mentioned above, but also with the space between a pair of infinitely long parallels, or between the arms of plane angle, and with much besides. In rational psychology too, we shall call a power of knowledge infinite (though short of omniscience) as soon as it is able to envisage an infinite set of truths, say those enunciating in turn the infinitely many digits in the decimal representation of the single quantity $\sqrt{2}$.

4. The commonest formulation of the infinitely great is: 'that which is greater than any *assignable* quantity.' The first necessity here is to determine more exactly what is meant by '*assignable*.' Is it to signify no more than that something is *possible* (that is, *capable* of actuality) or to signify no more than that it is *non-self-contradictory*? In the *former* alternative, we confine the idea of the *finite* exclusively to the category of things *actual*—either actual at all times, or actual at certain past times, or actual at certain future times, or at least *capable* of attaining actuality at some time or other. This seems in fact to be the meaning attached to the infinite by *Fries* (*Naturphilosophie*, §47) when he calls it the *uncompletable*. Yet linguistic custom applies the two notions, of the finite and of the infinite, not only to objects

enjoying actuality (as above all to God) but to others in whose case we cannot so much as speak of existence. Of this kind are the 'absolute propositions and truths' together with their components, the 'absolute presentations' (Vorstellungen an sich): for we assume both finite and infinite sets thereof. In the latter alternative, however, when we understand by the assignable everything barely *non-self-contradictory*, then we insinuate
13 the non-existence of the infinite into its very definition, since a quantity supposed greater than any non-self-contradictory quantity would have to be greater than itself, which is of course absurd.

Now in reality there is a third meaning in which the word '*assignable*' could be taken, to wit: as applying to all and only those things which can *in some way or other be given for us*, i.e. become the objects of *our experience*. Here I must really ask people whether they do not in every case use the words *finite* and *infinite* in such a sense (and whether their useful employment in science do not make it necessary to use them in such a sense) that they refer to a definite *internal* property of the thing so described, and by no means exclusively to a relation they bear to our *power of knowledge*, and still less to a relation they bear to our *sensitive faculties*—and this, quite irrespectively of whether we are able or unable to collect *experience* of them. As a consequence, the question whether a given object be finite or infinite certainly cannot depend on whether its quantity does or does not fall within our perception, or on whether we are able or unable to command a view of it.

§13

Now that we are agreed on the idea to be attached to the word *infinite*, and now that we have made ourselves clearly aware of the elements composing that idea, the next question to be asked is that concerning its *objective existence*—that is, whether there exist objects to which it can be applied, whether there exist sets which we may judge to be infinite in the sense here declared. This I venture to answer with a decided *affirmative*. Even in the *realm of things which do not claim actuality, and do not even claim possibility*, there exist beyond dispute sets which are
14 infinite. *The set of all 'absolute propositions and truths'* is easily seen to be infinite. For if we fix our attention upon any truth taken

at random, say the proposition that there exist such things as truths, or any other proposition, and label it *A*, we find that the proposition conveyed by the words '*A is true*' is distinct from the proposition *A* itself, since it has the complete 'proposition *A*' for its own subject. Now by the same law which enabled us to derive from the proposition *A* another and different one, which we shall call *B*, we are further enabled to derive a third proposition *C* from *B*, and so forth without end. The aggregate of all these propositions, every one of which is related to its predecessor by having the latter for its own subject, and the latter's truth for its own assertion, comprises a set of members (each member a proposition) which is greater than any particular finite set. The reader does not need to be reminded of the similarity borne by this series together with its principle of construction to the *series of numbers* considered in §8. The similarity consists in the fact that to each member of the former there corresponds a member of the latter; in the fact that howsoever large an integer be chosen, there exists a set of just so many among the above propositions; and finally in the fact that we can always continue the construction of such propositions—or rather, that such further propositions exist whether we construct them or not. Whence it follows that the aggregate of all the above propositions enjoys a multiplicity surpassing every individual integer, and is therefore infinite.

§14

For all the simplicity and clearness of the demonstration just
proffered, a large number of acute and learned men hold the 15
thesis itself to be not merely paradoxical, but even false in fact. They deny the *existence of anything infinite*. To nothing whatever—neither among the things possessing actuality, nor among the rest, neither to an individual nor to any aggregate—is it possible, according to their assertion, to attribute in any respect whatever an infinite set of members. The arguments which they bring against anything infinite in the realm of the actual will be considered later, for only later shall we bring our own arguments for the existence of such an infinite. Let us examine at the present moment, therefore, only the grounds upon which it is proposed to show that no infinite exists even among the things which lay no claim to actuality.

Firstly: 'Nowhere can an infinite set exist,' they say, 'for the simple reason that an infinite set can *never be united to form a whole, never be collected together in thought.*' I must stigmatise this assertion as a mistake, and as a mistake engendered by the false opinion that a whole consisting of certain objects *a*, *b*, *c*, *d*, . . . cannot be constructed in thought without previously forming separate mental representations (Vorstellungen) of its separate component objects. This is by no means true. I can think of the set, of the aggregate, or if you prefer it, the totality of the inhabitants of Prague or of Pekin without forming a separate representation of each separate inhabitant. So indeed do I act every time that I speak of this set and put forward my estimate, in the case of Prague, that the population lie between 100,000 and 120,000. As soon, in fact, as we possess a representation *A* which represents the objects *a*, *b*, *c*, *d*, . . . and no others, it is
16 extremely easy to arrive at a representation which represents the *aggregate* of all these objects taken together. Nothing more is needed than to combine the idea denoted by the word *aggregate*, and the notion *A*, in the manner expressed by the words '*the aggregate of all A.*' This single remark, whose correctness I trust will be evident to all, removes all the difficulties made against the idea of a set comprising infinitely many members: provided only that a specific concept is present, under which every member, and no non-member, can be subsumed. This is fulfilled for *the set of all 'absolute propositions and truths,'* the requisite specific concept being no other than the one already before us, to wit: 'an absolute proposition or truth.'

Nor ought I to let a *second* error in this first objection pass unreprehended. It is the opinion that 'sets do not exist unless there be somebody engaged in *thinking* of them.' In order to achieve a maximum of consistency in error, those who make this assertion should not only deny that there are *infinitely many* 'absolute propositions and truths,' but also deny that there be any 'absolute propositions and truths' *at all*. For if we have worked our way to a clear idea of 'absolute propositions and truths' and entertain no doubt of their objective existence, we shall not easily stumble into assertions like the one just quoted; or if we do stumble into them, we shall not easily persist in them. To the end that everybody may see this clearly, I take leave to raise the question whether there be no fluid or solid bodies at

the poles of the earth, no air or water or stones and the like, whether these bodies do not act upon one another according to fixed laws, in such wise for example that the velocities communicated at their collision are inversely proportional to their 17
masses, and whether all this happens even when no human or other thinking being is present to observe it? If this is granted—and who is not obliged to grant it?—then there exist 'absolute propositions and truths' which record these events completely, without there being anyone to think about them or to know them. In these propositions, moreover, there is frequent mention of wholes and sets: for every physical body is a whole, and produces very many of its effects only through the multitude of parts of which it is composed. Consequently, there exist wholes and sets even in the absence of a being to think of them. Were this not so, were these sets not objectively present, how could the judgements we pass upon them be true? Or rather, what content could these judgements have, if the presence of an observer of these processes were a necessary condition of the truth of these judgements? If I say: 'This boulder came loose from that rock before my very eyes, and hurtled down, cleaving the air as it fell,' then what I am supposed to mean must be something like this: 'As a result of my combining in my imagination certain simple substances at such a spot and such a height, there was formed a composite entity called by me a boulder; and this composite entity moved away from certain other simple substances which, as a result of my combining them too in my imagination, coalesced into a whole called by me a rock, and so on and so forth.'

Secondly: Someone may urge the following: 'Whether or no we choose to unite certain simple objects into an aggregate, it is only an operation *on our part*, and in most cases a very arbitrary operation; and relations arise between them only after we have carried out the operation. The midmost atom in this coat-button of mine, and the midmost atom in yonder steeple-knob, have absolutely nothing to do, and no connexion, with one another; only by my passing act of thinking them in combination is some sort of connexion established between them.'

This too I am obliged to contradict. Even before a thinking 18
being conjoined its mental representations, the two atoms were already exerting an influence upon one another, for example,

by their force of attraction; and except in the event of that thinking being proceeding in consequence of his thoughts to actions which do bring about a change in the relationships of the two atoms, it is altogether untrue that his thinking of them in combination be the sole cause of relations arising between them which otherwise would not obtain. If I am to judge correctly that this atom be the lower and that atom the higher, and judge correctly that as a result the former be attracted a little upwards by the latter, then all these things must take place even did I not think of them. And so forth.

Thirdly: others again would say: 'For the subsistence of an aggregate it is not requisite that it be *actually in the mind* of some thinking being, but it is necessary that it be *potentially* there. And since it is impossible for a being to exist, which is capable of representing to itself each separate one of an infinite set of objects, and then of combining these representations together, it must be impossible to have an aggregate embracing as members an infinite set of objects.'

Our answer to the first objection has already evinced the falseness of the hypothesis, here several times repeated, that the thought of an aggregate calls for a separate thought of each separate member. Nor are we under the necessity of appealing to an omniscient being as one for whom the distributive apprehension of an infinite set of objects causes no difficulty. For quite apart from these points, we cannot permit ourselves to concede even the first hypothesis, namely that the capacity of *becoming an object of thought* could be a necessary condition for the subsistence of an aggregate. For nothing can derive its capacity for existing
19 from its *capacity for being thought of.* On the very contrary, its capacity for existing is the prior ground upon which a rational being (in so far as not involved at the time in error) deems a thing *possible*, or as we somewhat inappropriately say, *thinkable.* The reader will become still more firmly persuaded of the correctness of this remark, and of the complete untenableness of the very widespread opinion I am here combating, if he tries to make clear to himself the elements which go to compose this extremely important concept of 'possibility.' To say that those things are *possible* which *can exist* is no genuine analysis of that concept; for that concept is contained entire in the meaning of the word '*can.*' Still more faulty would it be to set up as a

definition that just t h o s e things be possible *which can be thought of.* In the strict sense of the word, which includes mere *mental representation*, we can even 'think of' the impossible, and do really think of it as often as we make statements about it, declaring this or that to be impossible, for example, when we declare that there exists and could exist no quantity represented by 0 or $\sqrt{-1}$. Indeed, even if we exclude mere mental representation, and let 'thinking' only mean 'holding to be true,' it is still false that everything be possible which we can hold to be true. Error sometimes causes us to hold the impossible to be true, for example, to believe that we have squared the circle. The true formulation, which I anticipated above, must thus be as follows: 'Just those things are possible which a thinking being, if it judges in accordance with the truth, judges capable of existing,' in other words, judges to be 'possible'; and t h i s definition obviously contains a vicious circle! We are hence constrained, when defining the possible, to abandon altogether the reference to a thinking being, and to search for some other characteristic property. Sometimes we are told that the possible is *that which does not contradict itself.* Now it is true that every- 20
thing which contains a contradiction within itself is impossible, for example, that a sphere be not a sphere. On the other hand, however, not everything impossible is impossible in such wise that the contradiction lies between the components out of which its notion is put together. It is impossible, for example, to have a polyhedron with seven congruent faces—but the contradictoriness of this is not visible in the bare collocation of these words. We are thus forced to extend our definition. But were we to say that the impossible is that which stands in contradiction to any truth at all, we should be classifying as impossible everything which is non-existent—if only because the corresponding existential proposition would stand in contradiction to the truth of its non-existence. We should in this event make no sort of distinction between the actual and the possible, nor indeed between these and the necessary, whereas we all do make this distinction. From this we learn that the impossible does not contradict all truths indiscriminately, but only a definite category of truths; and it can now scarcely escape us, what category it is. It is the category of purely conceptual truths (reine Begriffswahrheiten). Whatever contradicts a purely conceptual

truth must be called an *impossibility*; and conversely, those things are possible which conflict with no purely conceptual truth. To anyone who has seen that this is the correct notion of possibility, it can hardly occur to assert that a thing only becomes possible by being actually thought of, that is, when some thinking being, without erring in his judgement, is looking upon it as possible. To say this would be to say as follows: 'For a proposition to contradict no purely conceptual truth, it is requisite that it should contradict no purely conceptual truth to have a thinking being which judges non-erroneously that the proposition contradicts no purely conceptual truth.' Who does not
21 see how entirely irrelevant is here the intervention of a thinking being? And once it is decided that the *act of thought* (das *Denken*) is not the source of possibility, what justification remains for denying the existence of infinite sets of objects on the alleged ground that they cannot be *assembled in thought*?

§15

The existence of infinite sets, at least with non-actual members, is something which I now regard as sufficiently proved and defended; as also, that the set of *all absolute truths* is an infinite set. Arguments like those in §13 will win assent for the statement that the set of *all numbers* is infinite—that is, the set of all the so-called natural or whole numbers, as defined in §8. Yet this statement also sounds *paradoxical*, and we may regard it as the *first paradox* to appear in the realm of mathematics—for the one just considered belongs, properly speaking, to a science more general than that of quantity.

'If each number,' it might be protested, 'is by definition a merely finite set, how can the set of *all* numbers be infinite? If we contemplate the series of the natural numbers

$$1, 2, 3, 4, 5, 6, \ldots,$$

we become aware that the numbers of the series lying between the first (unity) and any particular one of them form a set which is enumerated by that particular one. For example, those from one to six form a set of six numbers. Consequently, the set of *all* numbers must be enumerated by the *last* number, and being therefore itself a number, not be infinite.'

The illusiveness of this argument disappears at once if we

remember that in the natural series of natural numbers no term 22
occupies the *last place*, and that the notion of a last or highest number is an empty notion, being a self-contradictory one. In fact, the *principle of construction* of this series, as explained in §8, assigns to each of its terms a *following* term. This single remark must therefore be regarded as solving the present paradox.

§16

If the set of all *numbers* (namely the so-called *integers*) is infinite, then with all the greater certainty is the set of all quantities (as explained in §6 above, and in §87 of my *Wissenschaftslehre*) an *infinite* set. For according to that explanation, not only are all numbers also quantities, but there are far more quantities in existence than there are numbers: because the fractions $\frac{1}{2}$, $\frac{1}{3}$, $\frac{2}{3}$, $\frac{1}{4}$, . . . , and likewise the co-called *irrational* expressions $\sqrt{2}$, $\sqrt[3]{2}$, . . . , π, e, . . . denote quantities. Indeed, it is consistent with that explanation to speak of quantities *infinitely great* and others *infinitely small*, on condition that by an *infinitely great* quantity we only mean one which, after a unit has been chosen, presents itself as a whole of which every finite set of those units is only a part; and by an *infinitely small* quantity only one such that, the same unit being taken, this unit now presents itself as a whole of which every multiple of the proposed quantity is only a part. The set of all numbers manifests itself immediately as an indisputable example of an infinitely great *quantity*. I say advisedly, as an example of a quantity; and certainly not as an example of an infinitely great *number*; for this infinitely great multitude cannot, as we remarked in the previous paragraph, be given the name of number. On the other hand, if we take a
quantity which appears as *infinitely great* relative to a pre-selected 23
unit, and post-select that quantity as a differently chosen new unit, then a renewed comparison shows that the old unit realises the definition of the *infinitely small*.

§17

Space and *time*—which again do not belong to the domain of the actual, though they can be *determinations* of the actual—form a very important category of infinitely great quantities. Neither space nor time is an actual object, for neither is a *substance* and neither is a *quality* of a substance (Beschaffenheit) and neither

is found otherwise than as a determination of incomplete *substances*—of all incomplete substances, which means the same as: all limited, finite, dependent or created substances. Each of these, in fact, must always possess a definite position in space and in time, in such a manner that every simple substance must occupy a definite point in space at each definite *point in time.* Now the set of spatial points of which space is composed and the set of temporal points of which time is composed are both *infinite.* Nay more: not only is the set of all points in every region of space or of time an infinite set, but so too is the set of all the instants between two fixed instants howsoever proximate, or the set of all the spatial points between two fixed points howsoever proximate again. I have little need to involve myself in a defence of these propositions, because hardly any mathematician who
24 admits the infinite at all fails to admit this as well. To escape conceding so manifest an infinite as this one, the opponents of infinity *in any form* take refuge in the pretext that 'whereas we can admittedly always *think of more* instants or points than we have thought of so far, yet the set of those *actually* present remains constantly finite.' My answer to this is, that neither time not space, and therefore neither instants nor points, are anything actual at all; then in consequence it is absurd to speak of a finite set of them as *actually* existing; and even more absurd to imagine that these points receive their actuality as a result of our *thought.* For it would be deducible from this that the qualities of time and space depend on our acts of thought and judgement: and as a particular consequence, the ratio of the diameter of a circle to its circumference was rational as long as we erroneously believed it to be rational. Furthermore, all the properties of space which we are destined to discover in the future will only accrue to space when that future arrives!

Even if the adversaries are given a chance to correct their wording and say rather that only such thinking as corresponds with the truth goes to determine the true properties of time and space, the only outcome is a sheer tautology—namely that those things be true, which are true. From this, however, it is impossible to deduce anything at all against our assertion of the infinitude of time and space. At all events, it is silly (abgeschmackt) to say that time and space contain only so many points as we happen to be thinking of.

§18

Although the capacity of being regarded as a whole consisting of an infinite set of parts in some respect or other is a necessary condition for a quantity, or any other object, to be regarded by us in that same respect as infinite, yet it is not by itself suffi- 25
cient; for not every quantity which we consider as the sum of an infinite set of other quantities each separately finite need itself be infinite. It is universally recognised, for example, that irrational quantities such as $\sqrt{2}$ are finite relative to their standard unit, despite the fact that they can be looked upon as composed from an infinite set of fractions of the form

$$\frac{10}{14}+\frac{1}{100}+\frac{4}{1000}+\frac{2}{10000}+\ldots,$$

with integer numerators and denominators; as also that the sum of an infinite series of addends having the form

$$a+ae+ae^2+\ \ldots\ \text{in inf.}=\frac{a}{1-e}$$

amounts to the finite quantity $\frac{a}{1-e}$ whenever $e<1$ holds.* In

* Since the customary proof for the summation of this series appears to be not quite rigorous, perhaps I may be allowed the opportunity to make the following suggestions. Let us take $a=1$ and e positive, since the modification for other cases is obvious, and let us write the symbolical equation

$$S=1+e+e^2+\ \ldots\ \text{in inf.} \qquad (1)$$

Then this much is certain, that S denotes a positive quantity, be it finite or infinite. But for every integer value of n whatever, we have

$$S=1+e+e^2+\ldots+e^{n-1}+e^n+e^{n+1}+\ldots\ \text{in inf.}$$

or again,

$$S=\frac{1-e^n}{1-e}+e^n+e^{n+1}+\ldots\ \text{in inf.} \qquad (2)$$

for which we can also write

$$S=\frac{1-e^n}{1-e}+P' \qquad (3)$$

if we denote by P' the value of the infinite series

$$e^n+e^{n+1}+\ldots\ \text{in inf.},$$

whereby again this much is certain, that P' denotes a quantity which, 26
whether measurable or not, depends on e and n and is at any rate positive. But the same infinite series can be expressed as

$$e^n+e^{n+1}+\ldots\ \text{in inf.}=e^n(1+e+\ldots\ \text{in inf.}),$$

and the sum bracketed on the right side, consisting of infinitely many addends, now wears exactly the same appearance as the series put equal to S in the symbolic equation (1). Nevertheless, it cannot be regarded as

consequence of this, the assertion that a sum of infinitely many
27 separately finite quantities can nevertheless present itself as a merely finite quantity contains no contradiction; for if it did, we should not have found it possible to demonstrate its truth. The semblance of paradox which some may see in it originates

identical therewith, since the set of *its* addends and the set of those in (1), while agreeing in being infinite, fail to be identical in that the former has indisputably fewer terms than the latter. With full confidence, therefore, we can do no more than put

$$(1+e+e^2+\ldots \text{ in inf.})=S-P'',$$

whereby we may at any rate suppose that P'' is always a positive quantity dependent on n. We thus have

$$S=\frac{1-e^n}{1-e}+e^n(S-P'') \quad \ldots\ldots\ldots \quad (4)$$

or alternatively

$$S(1-e^n)=\frac{1-e^n}{1-e}-e^nP'',$$

and finally

$$S=\frac{1}{1-e}-\frac{e^n}{1-e^n}P'' \quad \ldots\ldots\ldots \quad (5)$$

A comparison of equations (3) and (5) yields

$$\frac{-e^n}{1-e}+P'=\frac{-e^n}{1-e^n}P'',$$

or alternatively

$$P'+\frac{e^n}{1-e^n}P''=+\frac{e^n}{1-e},$$

from which it can be inferred that if we choose n arbitrarily great and thus depress the value of

$$\frac{e^n}{1-e^n}$$

27 below any one [pre-]selected small quantity

$$\frac{1}{N}$$

however small this be, each of the two quantities

$$P' \text{ and } \frac{e^n}{1-e^n}P''$$

must separately sink below any one [pre-]selected value. This being so, each of the equations (3) and (5) indicates, in view of the fact that the value of S depends on e but not on n, that this

$$S=\frac{1}{1-e}$$

Note: In the translation of this footnote, unglossed and today perhaps disappointing save to historians, the dashed capitals P' and P'' have been used in place of the original capitals with *vertically* superscript 1 and 2. Moreover, P'' was misprinted on its second appearance as P', and in the equation second after (5) the last denominator carried an extraneous exponent n.

in forgetfulness of the fact that the addends become smaller and smaller. Nobody can be surprised if a sum of addends each of which halves its predecessor can never surpass the double of the initial addend, since at however late a term in the series we halt, exactly so much is wanting to make up that double, as the term in question has value.

§19

Even in the examples of the infinite so far considered, it could not escape our notice that not all infinite sets can be deemed equal *with respect to the multiplicity of their members.* On the contrary, many of them are *greater* (or *smaller*) than some other in the sense that the one includes the other as a part of itself (or stands to the other in the relation of part to whole). Many consider this as yet another *paradox*, and indeed, in the eyes of all who define the infinite as that which is incapable of increase, the idea of one infinite being greater than another must seem not merely paradoxical, but even downright *contradictory.* We on the other hand have now seen that this opinion rests on a notion of the infinite which is altogether discordant with linguistic usage. Our own definition—which agrees not only with 28
linguistic usage, but also with the purposes of scientific knowledge—does not tempt anyone to think it contradictory, or even astonishing, that one infinite set be greater than another. Who can fail to see that the length of the straight line taken from *a* unboundedly towards *R* and beyond is infinite? But who can

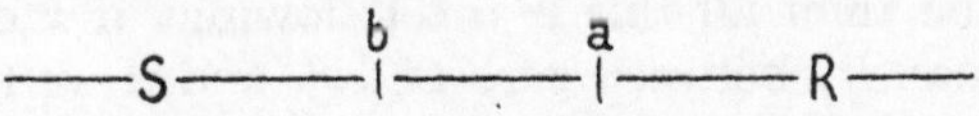

fail to see in addition, both that the unilaterally unbounded line from *b* rightwards exceeds that from *a* rightwards by the piece *ba*, and also that the bilaterally unbounded line taken both from *a* to *R* and beyond and from *a* to *S* and beyond, is greater still by an excess which is itself infinite? And so forth.

§20

We now pass on to consider a very remarkable peculiarity which can occur in the relations between two sets *when both are infinite.* Properly speaking, it always does occur—but, to the disadvantage of our insight into many a truth of metaphysics and physics and mathematics, it has hitherto been overlooked. Even now, when I come to state it, it will sound so paradoxical that we shall do well to linger somewhat over its investigation. My assertion runs as follows: When two sets are both infinite, they can stand in such a relation to one another that:

(1) it is possible to couple each member of the first set with some member of the second in such wise that, on the one hand, no member of either set fails to occur in one of the couples; and on the other hand, not one of them occurs in two or more of the couples; while at the same time,

(2) one of the two sets can comprise the other as a mere part
29 of itself, in such wise that the multiplicities to which they are reduced, when we regard all their members as interchangeable individuals, can stand in the most varied relationships to one another.

I shall conduct the proof of my assertion by means of two examples in which the above situation undeniably occurs.

Example I: let us choose any two abstract quantities, say 5 and 12. Then the set of all quantities between zero and 5 (or less than 5) is clearly infinite, as also the set of all quantities less than 12. With no less certainty is the latter set greater than the former, seeing that the former constitutes a mere part of the latter. We could even alter the quantities 5 and 12 into any others, and still be forced to admit that the corresponding two sets need by no means always stand in the same relation, but rather are able to enter into a great variety of relations. But no whit less true than all this is the following: if x denotes any arbitrary quantity between zero and 5, and if we fix the ratio between x and y by the equation

$$5y=12x,$$

then y is a quantity lying between zero and 12; and conversely, whenever y lies between zero and 12, x lies between zero and 5. The above equation also entails that to every value of x there belongs only a single value of y, and conversely. From these two premisses it follows that to every quantity x in the set between zero and 5 there corresponds a quantity y in the set between zero and 12 in such wise that, on the one hand, no constituent of either set remains uncoupled, and on the other hand, none appears in two or more of the couples.

EXAMPLE II: we intend to take a spatial object for our second 30
example. Those, to be sure, who already know that the qualities of space rest on those of time, and those of time on those of abstract numbers and quantities, need no example to show them that infinite sets like the above one in the realm of quantity must also exist in time and space. But for the sake of rightly conducting the application of our theorem in the sequel it is necessary all the same to study in detail at least one case in which sets of that kind exist. Let then *a*, *b*, *c* be three arbitrary points in a straight line, the ratio *ab* : *ac* being arbitrary save for the condition that *ac* shall be the greater of the two distances.

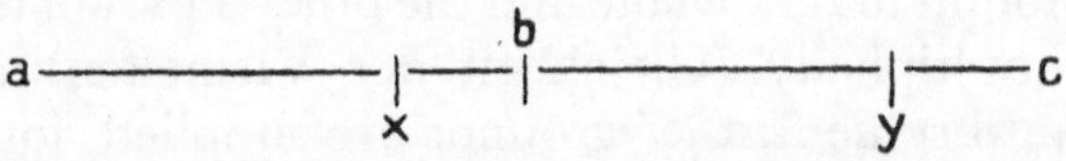

This being so, the set of points in *ab* and the set in *ac* will both be infinite, and this notwithstanding, the set of points in *ac* will surpass the set in *ab*, because the former contains all the latter together with all the points found in *bc* but not in *ab*. Nor can we escape the admission that arbitrary changes in the ratio *ab* : *ac* would produce great changes in the relations between the two sets. Yet with respect to the couples that can be formed with one term from each, the same can be said of this new pair of sets as was said of the previous pair of sets, in which the one consisted of all the quantities between zero and 5, and the other of all between zero and 12. For let *x* be any point in *ab*; and let *y* be specified, on *ax* produced, by means of the proportion

$$ab : ac = ax : ay.$$

Then *y* will be a point in *ac*. And conversely, if *y* is a point in *ac*, 31
and if *x* is specified from it by the same proportion, then *x* will be a point in *ab*. Moreover, different *x* will yield different *y*, and different *y* different *x*. From these two premisses it follows similarly that to every point in *ab* there corresponds one in *ac*, and to every point in *ac* one in *ab*, again in such a manner that we can say of the couples formed [in a fixed order] out of points thus corresponding, both that no point in either set fails to occur in some one couple, and that no point in either set makes two or more [homologous] appearances.

§21

The mere fact, therefore, that two sets A and B are so related that every member a of A corresponds by a fixed rule to some member b of B in such wise that the set of these couples $(a+b)$ contains every member of A or B once and only once, never justifies us, we now see, in inferring the *equality of the two sets, in the event of their being infinite*, with respect to the multiplicity of their members—that is, when we abstract from all individual differences. On the contrary, and in spite of their entering symmetrically into the above relation with one another, the two sets can still stand in a relation of inequality, in the sense that the one is found to be a whole and the other a part of that whole. This equimultiplicity (Gleichheit der Vielheiten) cannot be inferred until some further grounds are supplied, for example, an identical mode of specification (Bestimmungsgrund) or of generation (Entstehungsweise) for the two sets.

§22

32 As I am far from denying, an air of paradox clings to these assertions; but its sole origin is to be sought in the circumstance that the above and oft-mentioned relation between two sets, as specified in terms of couples, really does suffice, in the case of *finite* sets, to establish their perfect equi*multiplicity* in members. Whenever, in fact, two finite sets are constituted so that every object a in the one corresponds to another object b in the other which can be paired off with it, no object in either set being without a partner in the other, and no object occurring in more than one pair: then indeed are the two finite sets always equal in respect of multiplicity. The illusion is therefore created that this ought to hold when the sets are no longer finite, but infinite instead.

The illusion, I say—for a closer study reveals the fact that no such necessity exists, because the grounds upon which this holds for finite sets are bound up precisely with the finitude, and become inoperative for infinite sets. Suppose namely that both sets, A and B, are finite—or suppose something even less, yet still sufficient, namely that we have ascertained the finitude only of the one set A. Moreover, let us abstract from all individual differences, so as to compare the sets exclusively with respect to their multiplicities.

To begin, let us mark with 1 any arbitrary object in the set *A*,
with 2 any arbitrary other object, and generally, every following
object with the number of objects so far considered, the new one
included. Sooner or later we must come to a member of *A*, after
the marking of which we find in *A* no more objects to mark. This 33
much follows immediately from the idea of a finite or countable
multitude. If the above-mentioned last member of *A* was
marked *n*, the enumeral (Anzahl) of the members of *A* will be *n*.

Turning now to *B*, let us give to each object in *B* the same
mark as to its partner in *A*, according to our hypothesis as to the
couples. It must then turn out that the number of objects from
B which we have used up in this manner is likewise *n*: since each
of them received a mark showing how many had been used up
so far. This makes it clear that the number of objects in *B* is certainly not less than *n*; for one of them, namely the last to be
used up, actually does carry this number as its mark. On the
other hand, that number is also not greater than *n*; for if there
existed even a single one over and above those now used up, this
single one would lack a partner in *A*, contrary to the hypothesis.
Consequently, the number of objects in *B* is neither less than
nor greater than *n*, therefore equal to *n*. Both sets therefore have
one and the same multiplicity, or as we may also express it, an
equal multiplicity. This argument plainly fails so soon as the set
of objects in *A* is an *infinite* one; for in this case not only do we
who *do the counting* never arrive at a last term in *A*, but such a
last term is prevented from existing at all by the very force of the
definition of an infinite set. In other words, howsoever many
objects have hitherto been marked, there are still others yet unmarked. For this reason, the mere fact that *B* never runs short
of objects to couple with those of *A* offers no justification for
concluding that the multiplicities of the two sets are identical.

§23 34

What we have just said shows indeed that our oft-mentioned
relation between two sets entails their equimultiplicity when
they are finite, but *not when they are infinite*; but it omits to
explain how and why two sets that are infinite can still fail to
be equinumerous. For an explanation of this we must go back
to the adduced examples. They show us that the members *a* and
b which make up a couple *do not play exactly the same part in their*

respective sets. For if a' and b' form a second couple, and if we compare the relationships between a and a' as members of A with those of b and b' as members of B, we discover quite soon that they are different. Take the first example, and choose any two members of the set of quantities between zero and 5, say 3 and 4. Their partners in B are then clearly

$$\frac{12}{5}\times 3 \text{ and } \frac{12}{5}\times 4, \text{ that is, } 7\tfrac{1}{5} \text{ and } 9\tfrac{3}{5}.$$

If we mean by the *relationships* of two things, as we ought to mean, the aggregate of *all* the qualities manifest in their conjunction, then among all the relationships between 3 and 4 as members of the one set, and between $7\frac{1}{5}$ and $9\frac{3}{5}$ as members of the other, we do wrong to confine our attention exclusively to what is called *geometrical ratio*. We should pay heed to everything that belongs hither, in particular to the *arithmetical differences*. That between 3 and 4, which is 1, differs entirely from that between $7\frac{1}{5}$ and $9\frac{3}{5}$, which is $2\frac{2}{5}$. Thus, although every quantity in A or B allows of coupling with one and only one in
35 B or A, yet the set of quantities in B is other and greater than in A, since the *distance* between two quantities in B is other and greater than the *distance* between the corresponding quantities in A. One natural consequence is, that two quantities in B include between them a set of quantities other and greater than the set of quantities included between their partners in A. No wonder, therefore, that the *whole* set of quantities in B is other and greater than the set of quantities in A.

In the second example the situation is altogether similar. We intend, therefore, to say no more about it than this: that the points in ab which are partnered with points in ac lie uniformly *closer* to one another than do their respective partners in ac, because the distances between the former are to the distances between the latter in the ratio $ab : ac$.

§24

Now that the theorem in §20 can be considered as sufficiently proved and explained, one of the next deductions from it is this: *When two quantities are sums of infinite sets of addends equal in corresponding pairs, we are not entitled for that sole reason to equate them*; but only if we can first convince ourselves that the infinite multitude of those addends is identical in the two sums. It is not

disputed, to be sure, that addends determine their sum, and that equal addends yield equal sums. This holds not only for finite but also for infinite sets of summands. In the case of the latter, however, it is necessary to make sure that the infinite set of summands in the one sum really is identical with the infinite set of summands in the other sum; seeing, namely, that there are different kinds of infinite set. And to make sure of that point, we see from our theorem how altogether insufficient it is to be able 36
to pair off the terms in the one sum with those in the other. The conclusion will be unsafe unless *the two sets have identical terms of specification* (gleiche Bestimmungsgründe). The sequel will bring many examples of the absurdities in which a calculation with the infinite involves us if we fail to pay attention to this point.

§25

I now proceed to the assertion that there exists an infinite *even in the realm of the actual*, and not merely among the things which make no claim to actuality. Anyone who had arrived at the momentous conviction (whether by a chain of reasoning from purely conceptual truths or otherwise) *that there exists a God*, a Being whose existence is grounded in that of no other being, and precisely for this reason is a *universally perfect Being*, uniting in himself all powers and perfections which are compatible with one another at all, and each of them in the highest degree of which it is capable—such a person, I say, agrees by this very fact upon the existence of a Being possessed of infinitude in more than one respect; with respect to his *knowledge*, in that he *knows infinitely much*, to wit, the sum of all truths; to his *volition*, in that he *wills infinitely much*, to wit, the sum of every single possible good; and to his might, or *action ad extra*, in that he *confers actuality*, in virtue of his power of action ad extra, to *everything that he wills*. From this last attribute of God follows the existence of beings other than God, *creatures*, which we contrast with him and call merely *finite beings*, but in which for all that many a trace of infinitude can be found. For the *set* of such beings must 37
already be an infinite one, as also the set of all the *conditions* experienced by any single one of them during no matter how short an interval of time—because every such interval contains infinitely many instants. We therefore encounter infinites even in the realm of the actual.

§26

Even among well-instructed people who would not think of rejecting an infinite among things possessed of no actuality, such as 'absolute propositions and truths,' there are several who refuse assent to my last thesis. In their eyes, an infinite in the realm of the actual is disallowed by the immemorial *principle of the universal determinateness of the actual.* As against this I believe that I have already shown in my *Wissenschaftslehre* (Volume I, §45) that this principle is true of the non-actual in just the same sense in which it is true of the actual. It holds in all cases solely in the sense that, given any one object whatever, and any two *contradictory* qualities, one of the qualities must be affirmable of that object, and the other quality deniable of it. If it really conflicted with this principle to assume an infinite in the realm of the actual, then it would not be lawful to speak of an infinite among the unactual objects of our reflexion, and not lawful to admit an infinite set of absolute truths or of abstract numbers. But by pronouncing an object to be infinite, we are still very far from violating the principle in question. We only say, after all, that the object exhibits in some respect or other a multiplicity of parts or members which is greater than any particular num-
38 ber whatever—admittedly, therefore, a multiplicity which *does not allow of determination by an abstract number* (durch eine blosse Zahl). It does not follow from this at all that the multiplicity *cannot be determined in any manner whatever.* Nor does it at all follow that a single pair of opposite qualities *b* and non-*b* exist, both of which would have to be denied of it. A thing insusceptible of colour obviously cannot be specified by its colour, nor a thing incapable of emitting sound be specified by its note, and so forth; yet such things are by no means indeterminable, and constitute no exception to the rule that of the two predicates *b* and non-*b* (blue and non-blue, euphonious and non-euphonious) one must be attributable to each chosen object [and the other unattributable]—provided only that we interpret them in the manner necessary for them to remain contradictory. Just as 'non-blue' and 'non-fragrant' are predicable, though to be sure very remotely, of the Theorem of Pythagoras, exactly so is the bare statement that 'the set of points between *m* and *n* is infinite' one of the things predicable of this set. Frequently enough, a

very few statements are sufficient to determine such an infinite set of objects *completely*, i.e. in such a manner that *all* its attributes are deducible from the few actually named. Thus, we have completely determined the last-mentioned set of points between *m* and *n* the moment we determine the two points *m* and *n* themselves, say by an intuition (Anschauung) referring to them. For these few words suffice to decide with certainty whether any other proposed point does or does not belong to this set.

§27

Hitherto, I have had reason to defend the acceptance of an infinite against unjustified opposition; but now I must acknowledge with equal candour that many of the learned, especially in the ranks of the mathematicians, have gone too far in the other direction and accepted sometimes an infinitely great, sometimes an infinitely small, in cases where it is my firmest persuasion that none exists. 39

1. If an *infinite interval of time* were understood as one which had no beginning, or had no termination, or had neither beginning nor termination, and thus meant the whole of time or the aggregate of all instants, then I should have no objection to raise. At the same time, I hold it necessary to think of the *quantitative ratio* between two intervals of time, each of which lies between terminal instants, as a merely finite quantitative ratio, and determinable completely in terms of pure concepts. I would never allow the hypothesis that one duration, enclosed between terminal instants, be infinitely greater or less than another such duration. Yet too many mathematicians, as is well known, sometimes do this by speaking of infinitely great intervals of time which are nevertheless terminated at both ends, but more often by speaking of *infinitely small intervals of time* in comparison with which every finite interval of time, such as a second, is supposed to have to be avowed infinitely great.

2. The same thing happens with the *distances between pairs of points in space*, which in my opinion always bear a merely finite ratio to one another, a ratio determinable completely in terms of pure concepts; whereas nothing is more common among our mathematicians than to speak *of infinitely great* and *infinitely small distances*. 40

3. And finally, the same thing happens again with the *forces*

which both metaphysics and physics assume to act in the universe. We ought to suppose of them too, that none is ever infinitely greater or smaller than another, and that all of them stand in ratios determinable completely in terms of pure concepts; though, to be sure, we quite often permit ourselves to act otherwise. I shall not be able, of course, to explain the grounds of all these assertions on the present occasion to anyone who is still wholly unaware of what meaning I give to the words '*Intuition*' (Anschauung) and '*Concept*' (Begriff), the '*Derivability* of one proposition from another,' the '*Objective entailment*' (objektive Abfolge) 'of one truth by another' and of what *definition* I lay down for Time and Space. However, the following demonstration will not be altogether incomprehensible to those who have read at least the two treatises '*An Essay towards an objective Foundation of the Doctrine of the Composition of Forces*'* and '*An Essay towards an objective Foundation of the Doctrine of the three Dimensions of Space.*' †

The definitions of space and time lead immediately to the conclusion that all *dependent* (that is, created) substances continually act upon one another; as also to the conclusion that, given any two instants α and β, however near to one another or however far apart, it is permissible to regard the condition of the universe at α as a *cause*, α being the earlier instant; and the condition of the universe at the later instant β as an *effect*, if only as a mediate effect—provided always that any immediate interventions of God taking place in the interval αβ are reckoned with the causes. A further deduction is, that given the two
41 instants α and β, given all the *forces* with which created substances are endowed at the moment α, given the *position* each occupies in space, and given finally what divine interventions take place during αβ, we can infer the *forces* with which these substances are endowed at the moment β, together with the *positions* they then occupy, in just the same way as an *effect*, whether mediate or immediate, must be deducible from its complete cause. This necessitates in turn that all the qualities of the

* Prag 1842, by commission with Kronberger and Rziwnas. Note: The original title is: 'Versuch einer objektiven Begründung der Lehre von der Zusammensetzung der Kräfte.'

† Prag 1843, by commission with Kronberger and Rziwnas. Note: The original title is: 'Versuch einer objektiven Begründung der Lehre von den drei Dimensionen des Raumes.'

effect should permit of derivation from those of the cause by means of a major premiss whose terms are pure concepts, and which takes the form: 'Every cause with the qualities u, u', u'', . . . has an effect with the qualities w, w', w'', . . .'; and a simple corollary of this, which we need for our purpose, is that 'Every particularity (Umstand) in the cause, under whose variation the effect is not invariant, must *allow of complete determination* by means of pure concepts taken exclusively from among those required for the determination of the effect.'

After these preliminary remarks, our previous assertions are easily established [in corresponding paragraphs as follows].

1. Did but two instants α and β exist, whose distance were infinitely greater or smaller than the distance of other two γ and δ, we should be faced with the absurd result that the condition of the universe at β would be utterly impossible to determine from its condition at α together with a knowledge of the current divine interventions and the length of the duration $\alpha\beta$. The laying down of a standard unit of time is necessary, in fact, for the determination of the condition in which created beings are placed at a given moment, and even for the sole determination of the *magnitude of their forces*; for since these forces are nothing but *forces for the effecting of change* (Veränderungskräfte) 42
their magnitude can only be judged by observing the changes they bring about in a given stretch of time. Now take, as we must be allowed to take, the time interval $\gamma\delta$ as the unit. Then even in the most favourable case of all—that in which all the forces with which created substances are endowed at the instant α can be determined in terms of this unit, and in which everything else belonging to the complete cause of the condition of the universe at the instant β can be exactly determined as well—yet the distance between this instant itself and the instant α could not be expressed in terms of the chosen unit, inasmuch as it would turn out to be simply infinitely great or infinitely small. *Conversely*, if it is to be possible to consider every particular condition of the universe as the cause of every particular later condition, under the conditions now repeatedly mentioned, then we cannot have two instants α and β whose distance would turn out to be infinitely great or small in comparison with the distance of other two instants γ and δ.

2. Did but two points *a* and *b* exist in space, whose distance

were infinitely great or small in comparison with that of other two points c and d, then the determination of the condition of the universe at a specified instant α would require among other things the magnitude of the force of attraction or repulsion to be known, which the substance placed at a then exercises on the substance placed at b. But take, as is certainly permissible, the distance cd as the unit of length. Then even in the most favourable case of all—that in which we were successful with all other forces—it would still be impossible to succeed with this par-
43 ticular force. For even if, nay precisely because, the force of attraction exercised over the chosen unit of length cd by the substance A on the substance B (or on one otherwise completely similar to it) is a completely determinate quantity, the magnitude of that force would become indeterminate if the ratio $ab : cd$, on which it certainly also depends, were infinite and consequently itself indeterminate.

3. Finally: did but a single force k present itself as infinitely great or small in comparison with another force l, and this at an instant we shall label α, then even in the most favourable case—that in which all other forces proved finite when measured by their respective units of time and space, and in which, therefore, l too would be finite—the quantity k would still turn out to be infinitely great or small, that is, indeterminate. On this ground, however, the total situation of the universe at the instant α would come out indeterminate, and we should be faced with the impossibility of tracing any subsequent situation of the universe as an effect produced by the situation first mentioned.

§28

I am persuaded that the foregoing paragraphs lay down the fundamental rules for judging all the strange-sounding doctrines which have to be advanced in the sequel, the rules by which it must be decided whether they be errors due for abandonment, or theorems due for retention because they are true in spite of their appearance of preposterousness. The sequence in which we set forth these paradoxes shall be settled by what sphere of knowledge they belong to, and by their greater or less importance.

The first and most comprehensive science in which we encounter paradoxes of the infinite is the *general theory of quantity*,

as several examples have already taught us, and such paradoxes 44
are not wanting even in the *doctrine of number* itself. We intend in consequence to begin with these.

I confess that the mere *idea* of a *calculation with the infinite* has the appearance of contradicting itself. For to try to *calculate* anything means, after all, to attempt a *determination of it* in terms of number. But how can we possibly hope to determine the infinite by means of number—that very infinite which by our own definition must admit of being conceived as a set with infinitely many members, a set greater than any particular number and thus incapable of determination by a merely numerical datum? But this scruple vanishes when we reflect that a correctly conducted calculation with the infinite is not a numerical determination of what is therein not numerically determinable (namely, not a numerical determination of the infinite multitude as such) but only aims at determining the ratio [or relationship (5)] between one infinite and another; a thing which can, in fact, be done in certain cases, as we hope to show by several examples.

§29

All who admit the existence of multitudes that are infinite, and hence also of infinite quantities, must go on to admit that infinite quantities exist which exhibit manifold differences with respect to their actual magnitude or bigness.(6) If the sequence of the natural numbers is set down in the form

$$1, 2, 3, 4, \ldots, n, n+1, \ldots \text{ in inf.},$$

then the symbol (Zeichnung)

$$1+2+3+4+ \ldots +n+(n+1)+ \ldots \text{ in inf.}$$

will represent (darbieten) the *sum* of these natural numbers; 45
and the following symbol

$$1^0+2^0+3^0+4^0+ \ldots +n^0+(n+1)^0+ \ldots \text{ in inf.},$$

whose addends are all simple unities, will represent the bare *set* of all the natural numbers. If we denote (7) the latter by N_0, construct the merely symbolic equation

$$1^0+2^0+3^0+4^0+ \ldots +n^0+(n+1)^0+ \ldots \text{ in inf.} = N_0 \quad (1)$$

and denote the set of natural numbers from $(n+1)$ onwards similarly by N_n, constructing the similar equation

$$(n+1)^0+(n+2)^0+(n+3)^0+ \ldots \text{ in inf.} = N_n, \quad (2)$$

subtraction yields us the quite irreproachable equation

$$1^0+2^0+3^0+4^0+\ .\ .\ .\ +n^0=n=N_0-N_n,\ .\ .\ (3)$$

whence we learn how two infinite quantities N_0 and N_n can sometimes have an altogether definite finite difference.

If on the contrary we denote the quantity representing the *sum* of all the natural numbers by S_0, or write down the merely symbolic equation

$$1+2+3+4+\ .\ .\ .\ +n+(n+1)+\ .\ .\ .\ \text{in inf.}=S_0, \quad (4)$$

we shall no doubt apprehend at once that S_0 must be far greater than N_0; but it will be less easy for us exactly to ascertain the difference between these two infinite quantities, or even their (geometrical) *ratio* to one another. For if we tried to do what many indeed have tried to do, and set down the equation

$$S_0=\frac{N_0(N_0+1)}{2},$$

46 we could justify it on scarcely any other grounds than that the equation

$$1+2+3+4+\ .\ .\ .\ +n=\frac{n(n+1)}{2}$$

holds for every finite set of terms, whence it appears to follow that, when we pass to the whole infinite set of numbers, n simply becomes N_0. But that is not the case, for it is absurd to speak of an infinite series as having a last term with the value N_0.

The merely symbolic equation (4) [*sic*] having meanwhile been laid down, it will indeed be permissible to derive the following two equations from it by letting N_0 multiply each side:

$$1^0N_0+2^0N_0+3^0N_0+\ .\ .\ .\ \text{in inf.}=N_0^2,$$
$$1^0N_0^2+2^0N_0^2+3^0N_0^2+\ .\ .\ .\ \text{in inf.}=N_0^3,$$

and so forth, which convinces us that there exist infinite quantities of so-called *higher orders* as well, such that one exceeds the other infinitely many times. The existence, moreover, of infinite quantities bearing to one another any prescribed ratio, rational or irrational, say $\alpha:\beta$, already follows from the fact that if N_0 only denotes any constant infinite quantity, then αN_0 and βN_0 are a pair of quantities which are also infinite and bear to one another the prescribed ratio $\alpha:\beta$.

It will presumably be not less evident that the entire *set* (or multiplicity) of quantities lying between two given ones, say 7 and 8, depends only on the distance 8–7 between those terminals and is hence equal (gleich) and necessarily equal to any other for which that distance is equal—and this, despite the fact 47
that it is an *infinite* set and thus incapable of determination by any number howsoever great. On this supposition, and denoting the set of all quantities lying between *a* and *b* by

$$\text{mult}(b-a),$$

there must be numberless equations of the form

$$\text{mult}(8-7)=\text{mult}(13-12),$$

as also of the form

$$\text{mult}(b-a):\text{mult}(d-c)=(b-a):(d-c),$$

against whose validity there is no sound objection.

§30

Now that the possibility of *calculating with the infinitely great* has been vindicated by these few examples, we assert the like for the *infinitely small*. For if N_0 is infinitely great,

$$\frac{1}{N_0}$$

necessarily represents an infinitely small quantity, and we shall have no reason for denying objective reference to such an idea, at any rate in the *general* theory of quantity. To give a single example: if the probability be asked that anyone should shoot off a bullet at random, and yet have its midpoint pass exactly through the midpoint of yonder apple on yonder tree, then all must agree that the set of all the possible cases, with their various degrees of probability, is infinite: whence it follows that the degree of probability in question has a value equal to or less than $1/\infty$. This alone suffices to show that we have infinitely many infinitely small quantities, with one standing 48
to another in any prescribed ratio, and in particular, such that one be infinitely greater than another; as also that among the infinitely small quantities we have, just as among the infinitely great ones, an infinite number of different orders. Furthermore, if certain rules are kept, it will indeed be possible

to find very many correct equations between quantities of this sort.

Suppose we have ascertained, for example, that the value of a variable quantity y depends upon another x in such wise that the equation

$$y=x^4+ax^3+bx^2+cx+d$$

constantly holds between them, and that it is consistent with the nature of the particular class of quantities to which x and y belong that they should become infinitely small and thus be susceptible of infinitely small increments. Then, if we let x increase by the infinitely small amount denoted by dx, and denote by dy the change which y then undergoes, the following equation necessarily holds:

$$y+dy=(x+dx)^4+a(x+dx)^3+b(x+dx)^2+c(x+dx)+d,$$

which incontestibly entails the following:

$$\frac{dy}{dx}=(4x^3+3ax^2+2bx+c)+(6x^2+3ax+b)dx+(4x+a)dx^2+dx^3,$$

and this in turn exhibits the two infinitely small quantities as depending not only on a, b, c and x, but in addition on the value of the variable dx.

§31

But the majority of the mathematicians who ventured on a calculation with the infinite went much farther than is permissible on the principles here laid down. To assume an
49 infinitely great and an infinitely small among quantities whose nature does not allow of them—and examples of which will be given later—was not the only liberty which they took without hesitation. They also made bold to declare equal to one another, or greater or less than one another, quantities arising from the summation of an infinite series, on the sole ground that the terms of the series were equal in pairs or unequal in pairs, and despite their obvious inequality when considered as sets. They had the hardihood to make the statement not only that every infinitely small quantity *vanishes like an ordinary zero* when added to a finite one, and not only that every one of a *higher* order does so in conjunction with one of *lower* order, but even that every infinitely great quantity of lower order does so in conjunction with another of *higher* order. And in order to find some justification for this method of calculation they had recourse to the

statement that it is permissible to divide by zero, and that the quotient

$$1/0$$

is in reality nothing else than an *infinitely great quantity*, and the quotient

$$0/0$$

a quantity *completely indeterminate*. We must show how false and misleading these ideas are, because they are more or less in vogue even today.

§32

As recently as 1830, a writer signing himself M. R. S. tried to prove in *Gergonne*'s *Annales de Mathématiques* (Volume 20, Number 12) that the well-known infinite series

$$a-a+a-a+a-a+\ldots \text{ in inf.}$$

has the value $a/2$. Setting its value equal to x, he believed himself entitled to deduce that 50

$$x=a-a+a-a+\ldots \text{ in inf.}=a-(a-a+a-a+\ldots \text{ in inf.}),$$

and that the bracketed series, being identical with that whose value is being sought, could again be set equal to x, yielding

$$x=a-x,$$

and consequently

$$x=a/2.$$

The fallacy does not lie deep. The bracketed series no longer has the same identical set of terms as the one originally put equal to x, because it wants the initial a. If it had a value at all, that value would have to be denoted by $x-a$, and this would give us the mere identity

$$x=a+x-a.$$

Some one may perhaps urge that 'there is something paradoxical in the thought that this series, which is assuredly not infinitely great, should have no exactly determinable and measurable value at all—and this all the more, in that it is generated by the infinitely continued division of a, by 2 in the form $1+1$; a fact which speaks for the correctness of the assumption that its true value is after all none other than $a/2$.'

In answer to this I recall first that there is nothing intrinsically impossible in our having *quantitative expressions* (Grössenausdrücke) which denote *no actual quantity* (keine wirkliche

Grösse bezeichnen), and of these, zero is and must be a recognised example.

In particular: once we declare that we intend to consider a *series* only as a quantity, viz. only as the *sum* of its terms, then in virtue of the *definition* of a sum (which classes it with sets,
51 therefore with those aggregates from the order of whose members we abstract) it must be such as to undergo no variation in its value howsoever we vary the sequence of its terms. Among quantities it is necessary to have

$$(A+B)+C=A+(B+C)=(A+C)+B.$$

This characteristic, however, is a clear proof to us that the symbol in question, to wit

$$a-a+a-a+a-a+ \ . \ . \ . \text{ in inf.}$$

is not the expression of an actual quantity. For if a quantity were really represented by it at all, that quantity would suffer no alteration when we changed the symbol to:

$$(a-a)+(a-a)+(a-a)+ \ . \ . \ . \text{ in inf.} \qquad (1)$$

because we should be doing nothing more than combine each consecutive pair of terms into a partial sum, a thing which must be possible, because the given series really has no *last* term. By this process, however, we obtain

$$0+0+0+ \ . \ . \ . \text{ in inf.,}$$

which can obviously be equal only to 0.

Again, if a quantity were really represented by our expression, it would suffer equally little alteration on being transformed as follows:

$$a+(-a+a)+(-a+a)+(-a+a)+ \ . \ . \ . \text{ in inf.,} \quad (2)$$

where, setting aside the first term, we combine into a partial sum each consecutive pair of following terms; or again on being transformed as follows:

$$-a+(a-a)+(a-a)+(a-a)+ \ . \ . \ . \text{ in inf.,} \quad (3)$$

which is what becomes of (1) when we transpose its terms in successive pairs and then effect the alteration which led from (1) to (2). Now if the expression under investigation were not *devoid of objective reference*, the symbols (1), (2) and (3) would have to denote one and the same quantity; because it is evident that the
52 mental representation (Vorstellung) of the sum of one and the

same set of quantities cannot represent several quantities *different from one another*, as does happen for example with the representations

$$\sqrt{(+1)}, \text{ arc sin } (\tfrac{1}{2}),$$

and many others. Even the quantitative representation

$$1-1+1-1+1-1+\ .\ .\ .\ \text{in inf.},$$

unless wholly devoid of objective reference, would have an equal right to be put equal to zero (which is usually called a quantity, though in an improper sense) and to $+a$ and to $-a$, which is thoroughly absurd, and entitles us to conclude that we actually do have before us a representation devoid of objective reference.

It is true, again, that the series under review does appear as the quotient in a continued division of a, by 2 in the form $1+1$; but none of the series generated in this fashion can yield the true value of the quotient (in our case $a/2$) unless perchance the remainders arising on further division become smaller than any chosen small quantity, the easily understood reason being that the division continually leaves a remainder (in our case alternately $-a$ and $+a$). The condition just mentioned was fulfilled by the series studied in §18, which is generated on division of a by $1-e$, provided $e<1$. But when $e=1$, as in the case now before us, and still more when $e>1$, which causes the remainders to increase at every step in the division process, nothing is easier to understand than that the value of the series cannot be equated to the quotient

$$\frac{a}{1-e}.$$

How, indeed, could we possibly put the series

$$1-10+100-1000+10000-100000+\ .\ .\ .\ \text{in inf.},$$

whose terms alternate in sign, and which is generated on 53
division of 1 by $1+10$, equal to $1/11$? And who, to crown it all, would care to estimate at $-1/9$ the value of the series

$$1+10+100+1000+10000+\ .\ .\ .\ \text{in inf.},$$

all of whose terms are positive, on the sole ground that the development of the fraction

$$\frac{1}{1-10}$$

leads to that series? Nevertheless, the abovementioned M. R. S.

undertakes to defend such summations, and regards the equation

$$1-2+4-8+16-32+64-128+\ .\ .\ .\ \text{in inf.}=\tfrac{1}{3}$$

justified on the sole ground that, as he alleges,

$$\begin{aligned} x&=1-2+4-8+16-32+64-\ .\ .\ . \\ &=1-2(1-2+4-8+16-32+\ .\ .\ .)=1-2x; \end{aligned}$$

whereby he fails to notice that the bracketed series is by no means identical with the one first taken, because it no longer contains the identical set of terms. The lack of an objective reference for this quantitative expression is manifested in the same way as for the one previously considered, namely by its leading to contradictory results. On the one hand, in fact, we must have

$$\begin{aligned} &1-2+4-8+16-32+64-\ .\ .\ . \\ &=1+(-2+4)+(-8+16)+(-32+64)+\ .\ .\ . \\ &=1+2+8+32+64+\ .\ .\ ., \end{aligned}$$

and on the other, with equal certainty,

$$\begin{aligned} &=(1-2)+(4-8)+(16-32)+(64-128)+\ .\ .\ . \\ &=-1-4-16-64-\ .\ .\ ., \end{aligned}$$

so that by two legitimate processes two values are found for the same expression, one of them infinitely great and *positive*, the other infinitely great and *negative*.

54 §33

In order not to go astray in our calculations with the infinite, we must therefore never allow ourselves, upon two infinite quantities arising from the summation of two infinite series, to put the first quantity equal to (or greater than, or less than) the other quantity, solely because each term in the one series is equal to (or greater than, or less than) some one corresponding term in the other series. Nor ought we even to declare the first sum *greater*, solely because it contains all the addends of the second sum together with others besides—no, not even if these excess terms are (all positive and) infinite in number. For despite all this it still remains possible that the first sum be smaller, or indeed infinitely smaller, than the second sum. One example is to hand in the well-known sum of the *squares* of all the natural numbers, compared with the sum of their *first powers*. No one, assuredly, can contest that every term in the series of *squares*

$$\left.\begin{array}{r} 1^2+2^2+3^2+4^2+5^2+6^2+7^2+8^2+9^2+10^2+\ldots \text{ in inf.} = \\ 1+4+9+16+25+36+49+64+81+100+\ldots \text{ in inf.} = \end{array}\right\} S_2,$$

being itself one of the natural numbers, also occurs in the series of first powers of the natural numbers

$$1+2+3+4+5+6+7+8+9+10+11+12+13+14+15 \\ +16+\ldots \text{ in inf.} = S_1;$$

or that the latter series S_1 contains all the terms of S_2 and many, nay infinitely many more which are wanting in S_2 because they are not square numbers. Nevertheless, the sum S_2 of the square numbers is not equal forsooth to S_1, the sum of the first powers but incontestably greater. For in the first place the *set of terms* 55
in the two series (not yet considered as sums at all, and hence not partitionable into arbitrary subsets of members) is the same, in spite of all appearances to the contrary. The mere raising to the second power of each separate term in S_1 certainly alters the quality and amount of these terms, but it does not alter their multitude. But if the set of terms in S_1 and S_2 is the same, then it lies open to view that S_2 must be much greater than S_1 in that apart from the *first* term in each series each subsequent term in S_2 is decidedly greater than the homologous term in S_1; in such a manner that, considered as a quantity, S_2 contains the whole of S_1 as a mere part of itself, nay, over and beyond this mere part, it contains as a second part yet another infinite series, equinumerous in terms with S_1, to wit:

$$0, 2, 6, 12, 20, 30, 42, 56, \ldots, n(n-1), \ldots \text{ in inf.},$$

in which, apart from the first *two* terms, all the following terms exceed their homologues in S_1, so that the sum of this whole new series is again incontestably greater than S_1. If then we subtract the series S_1 for the second time from the above remainder, we obtain as our *second* remainder a series equinumerous in terms:

$$-1, 0, 3, 8, 15, 24, 35, 48, \ldots, n(n-2), \ldots \text{ in inf.},$$

in which, apart from the first *three* terms, once again all the following terms exceed their homologues in S_1; so that this third [*sic*] remainder can without fear of contradiction be deemed greater than S_1. Now since this argument can be continued without end, it is clear that the sum S_2 is infinitely many times greater than the sum S_1, for in the general case (8):

56 $$S_2 - mS_1 = (1^2 - m) + (2^2 - 2m) + (3^2 - 3m) + (4^2 - 4m) + \ldots + (m^2 - m^2) + \ldots + n(n-m) + \ldots \text{ in inf.},$$

a series in which only a finite set of terms (the first $m-1$) are negative, the *m—th* is zero and all the following ones are positive and increase to infinity.

§34

We cannot cast an appropriate light on the incorrectness of the other assertions mentioned in §31 until we define the idea of *zero* somewhat more precisely than is customary.*

It is beyond dispute that all mathematicians desire to attach to the symbol 0 an idea of such a character that, when A is any quantitative expression whatever (irrespective of whether it corresponds to an actual quantity or lacks objective reference) the two equations

$$\text{I. } A - A = 0, \qquad \text{II. } A \pm 0 = A$$

can always be legitimately written down. Now everybody will admit at this point that this can only be done if we look upon the symbol 0 itself not as the representation of an actual quantity, but as the absence(9), the bare absence of any quantity; and upon the symbol $A \pm 0$ as a command, neither to add to nor to subtract from the quantity denoted by A. On the other hand, it would be an error to believe that the simple declaration 'zero is a quantitative representation devoid of objective reference' sufficiently determines the idea which mathematicians attach to
57 this symbol. For there exist other quantitative notations of common usage in mathematics, such as the now important symbol $\sqrt{-1}$ in analysis, which are equally devoid of objective reference, and which for all that are by no means to be regarded and treated as equivalent to 0. But if we define the meaning of the symbol 0 more precisely by requiring that the two equations I and II shall universally hold, then we set up a concept which

* I pay willing tribute to the merits of M. Ohm for being the first to draw the attention of the mathematical public to the difficulties inherent in the idea of zero, in his valuable work: *Versuch eines vollkommen consequenten Systems der Mathematik* (2. Auflage, Berlin 1828).

is, on the one hand, quite as wide as past usage and the interests of science demand, and on the other hand, restricted enough to prevent its abuse.

Upon closer scrutiny, the postulate that equations I and II be universally valid does more than just demarcate the idea of zero: it also confers upon the idea of *adding* and *subtracting* (which appear under the signs + and −) a peculiar kind of extension which redounds greatly to the advantage of science.

That same advantage of science requires in addition that we conceive of *multiplication* too in a sense so wide that, whatever A is (whether a finite quantity, an infinitely great quantity, an infinitely small quantity, or like $\sqrt{-1}$ merely a quantitative representation lacking objective reference, or finally zero itself) we may in all cases write down the equation

$$\text{III.}\quad 0\times A = A\times 0 = 0.$$

Lastly, and again in the interests of science, we must demand so wide a conception of *division* as we possibly can without coming into conflict with one of the previous three equations, and hence allow the symbol B in the equation 58

$$\text{IV.}\quad B\times\left(\frac{A}{B}\right)=\left(\frac{A}{B}\right)\times B = A$$

as wide a range as is ever compatible with the universal validity of those three equations. Now though they permit B to denote any arbitrary finite or infinitely great or infinitely small quantity, and even the imaginary $\sqrt{-1}$, they simply do not permit B to be put equal to 0, in other words, they do not permit us ever to use zero, or any expression equivalent to zero, as a *divisor*. For equation III requires $0\times A=0$ for (10) every A, and if we put $B=0$ in equation IV, we should have to have

$$B\left(\frac{A}{B}\right)=0,$$

and this would agree with the requirement in IV that

$$B\left(\frac{A}{B}\right)=A$$

only in the solitary case of $A=0$. Lest we fall into contradiction, therefore, we must lay down the rule *that zero, or an expression equivalent to zero, may never be used as a divisor in an equation which purports to be more than a mere identity*, such as

$$\frac{A}{0}=\frac{A}{0}.$$

The need for observing this rule is shown not only by what has just been said, but also by a large number of very absurd consequences which ensue from entirely correct premises as soon as we permit ourselves to divide by zero.

Let a be any real quantity whatever. Then the well-known and certainly quite correct method of division yields, so soon as we are allowed to divide by the expression $1-1$, which is equivalent to zero, the following equation:

$$\frac{a}{1-1}=a+a+\ .\ .\ .\ +a+\frac{a}{1-1},$$

where the number of summands a may be any we desire. If we now subtract the same quantitative expression $\frac{a}{1-1}$ from both
59 sides, then we obtain the highly absurd equation

$$a+a+\ .\ .\ .\ +a=0.$$

Again, if a and b are a pair of different quantities, the pair of identities

$$a-b=a-b,$$
$$b-a=b-a$$

will hold, and by their addition we obtain in turn

$$a-a=b-b,$$
$$a(1-1)=b(1-1).$$

Were it permissible, however, to divide both sides of an equation by a factor equivalent to zero, we should obtain the absurd result that $a=b$, irrespective of the values of a and b. And yet it is a matter of common knowledge how easy it is in longer calculations to hit upon a false result, if we remove a common factor from both sides of an equation without first ascertaining that it does not vanish.

§35

It is now easy to prove the incorrectness of the assertion made by so many people that, in the course of addition or

subtraction, both an infinitely small quantity of higher order when conjoined with another of lower order or with a finite quantity, and an infinitely large quantity of any order when conjoined with another of higher order, and a finite quantity when conjoined with an infinitely large one, *vanish after the manner of a mere zero*. If this is to be understood in the sense that, when M is infinitely greater than m, we can simply leave m out of the compound expression
$M \pm m$, even when M itself vanishes in the course of the 60
calculation (say by subtraction of another equal to it) then I have no need to begin demonstrating the erroneousness of this rule—and in the common expositions, moreover, which are in some degree even more careless than the phraseology I have used above, no effort is made to prevent such a misunderstanding.

The rejoinder will be made that such is not the true meaning of what was said. It will be urged that when the quantities M and $M \pm m$ are declared equal it is not suggested that they give the same result when entering into further additions and subtractions, but only that they give the same result on going through a process of measurement in terms of a quantity N of equal rank with them, that is, standing in a finite and completely determinate ratio to one of them, say to M. So much at the very least are we justified in expecting to find in the definition of the phrase that two quantities are *equal in magnitude*. But do M and $M \pm m$ fulfil even this requirement? [Let us see.] If one of them, say M, stands in an irrational ratio to the measure N, it can certainly turn out to be the case that (as in the commonest type of measurement) given any [whole] number q, however great, another [whole] number p can be found to render (11)

$$\frac{p}{q} < \frac{M}{N} < \frac{p+1}{q},$$

and it can so happen that $\frac{M \pm m}{N}$ constantly remains within the same bounds, in other words, that we also have

$$\frac{p}{q} < \frac{M \pm m}{N} < \frac{p+1}{q}.$$

But in the alternative case where the ratio $M:N$ is rational, there exists a number q for which, alongside

$$\frac{M}{N}=\frac{p}{q},$$

we have either

$$\frac{M\pm m}{N}<\frac{p}{q} \text{ or else } \frac{M\pm m}{N}>\frac{p}{q};$$

61 so that in this alternative case the difference between the two quantities betrays itself even through the medium of abstract *numbers* (finite quantities). What right, then, have we to call them equal?

§36

Following in the footsteps of *Euler*, a number of mathematicians sought to avoid such contradictions by taking refuge in the declaration that infinitely small quantities are in reality *mere zeros*, and infinitely great quantities are the quotients arising from finite quantities upon division by a mere zero. This declaration, it is true, more than justified the vanishing or the rejection of an infinitely small quantity as the addend to a finite augend; but it only increased the difficulty of making intelligible the existence of infinitely great quantities, the emergence of a finite quantity from the division of two infinitely small or of two infinitely great quantities, and the occurrence of infinitely great and small quantities of higher orders. For the infinitely great quantities make their appearance, on this view, as a result of division by zero or by a quantitative expression equivalent to zero (something devoid, strictly speaking, of objective reference) and hence in a manner forbidden by the laws of calculation; while all the finite or infinite quantities which were supposed to result from the division of one infinite quantity by another bore the stain of multiply illegitimate birth.

The most plausible plea for this calculation with zeros seems to be the method of computing a quantity y which depends on a variable x by means of the equation (12)

$$y=\frac{F(x)}{\Phi(x)}$$

62 in cases where a particular value $x=a$ reduces either the denominator alone, or both denominator and numerator, to zero. In the former alternative, when $\Phi(x)=0$ but $F(x)$ continues to be a finite quantity, people infer that y has become

infinitely great; and in the latter alternative, when both $\Phi(x)=0$ and $F(x)=0$, that the two expressions $\Phi(x)$ and $F(x)$ both contain the factor $(x-a)$ once or oftener, and hence have the form

$$\Phi(x)=(x-a)^m\varphi(x), \qquad F(x)=(x-a)^n f(x),$$

where $\varphi(x)$ or $f(x)$ may possibly denote a constant. If thereupon $m>n$, people infer that, even after cancellation of the factor common to numerator and denominator—which leaves the value of the fraction $F(x)/\Phi(x)$ unaltered—the denominator still becomes zero, and continue to argue that $x=a$ yields an infinitely great y. But if $m=n$, they regard the finite quantity expressed by $f(a)/\varphi(a)$ as the true value of y, on the grounds that $F(x)/\Phi(x)=f(x)/\varphi(x)$ must hold. Finally, if $m<n$, they argue from the fact that in this event

$$\frac{F(x)}{\Phi(x)}=\frac{(x-a)^{n-m}f(x)}{\varphi(x)}$$

becomes zero for $x=a$, that the choice $x=a$ makes y vanish.

My verdict on this process is as follows. When the value of y for $x=a$ is declared in various foregoing cases to be ∞ great, this can *only* be true if the quantity y belongs to a kind *capable* of becoming infinitely great; and even *then*, only by accident; for the given expression, here requiring division by zero, is not the source from which the result springs. This is the inescapable 63
fact. From the mere circumstance that the value of y is said always to be the one yielded by the expression $F(x)/\varphi(x)$, we can deduce the properties of the quantity y only for all those values of x which represent an actual quantity, but not for those which render the expression *devoid of objective reference*, as happens if its numerator, or even just its denominator, still more if its numerator and its denominator together, should become zero. We can indeed say that in the first-mentioned case, where $\varphi(x)$ alone vanishes, the quantity y becomes *greater* than any already given quantity; in the second case, where $F(x)$ alone vanishes, that y becomes *smaller* than any already given quantity; and lastly in the third case, where the fraction $F(x)/\varphi(x)$ has equally many factors $(x-a)$ in numerator and denominator, that y comes arbitrarily near to $f(a)/\varphi(a)$ as we push x arbitrarily near to the value a; but nothing follows from all this concerning the nature of this value at places where the expression $F(x)/\varphi(x)$ *loses objective reference* (in other words, represents no value at all)

because it then assumes either the form o itself, or the form *c*/o, or else even the form o/o. For the theorem that the value of a fraction is not altered by cancellation, though valid in all other cases, is invalid when the cancelled factor *vanishes*. Otherwise, and once we had a right to say that

$$\frac{2\times 0}{3\times 0}=\frac{2}{3},$$

we should have an equal right to say of any arbitrary quantity, say of the number 1000, that

$$1000=\frac{2}{3}.$$

For $3000\times 0=0$ is doubtlessly just as true as $2\times 0=0$. If we may write

$$\frac{2\times 0}{3\times 0}=\frac{2}{3},$$

then we may also write

64 $$\frac{2\times(3000\times 0)}{3\times(2\times 0)}=\frac{(2\times 3000)\times 0}{(3\times 2)\times 0}=\frac{2\times 3000}{3\times 2}=1000.$$

The fallacy which here is so evident is less so in the former case because we then divided by the zero-equivalent factor $(x-a)$ in a form which concealed the fact of its vanishing. And our confidence in being allowed to do so is increased by the circumstance that the process, allowable as it is in every other case, here yields precisely the value which we consider ourselves entitled to expect, namely: if it is *finite* at all, then just what the law of continuity demands; zero, if the neighbouring values decrease towards zero; and infinitely great if the neighbouring values increase to infinity. We are forgetting, however, that the law of continuity is far from being observed by all variable quantities. We are forgetting that a quantity which becomes arbitrarily small when we bring x arbitrarily close to a need not for that sole reason become zero for $x=a$, and that one which increases to infinity while x is approaching a has as little need to become really infinite for $x=a$. Numerous geometrical quantities know of no law of continuity, for example: the magnitudes of the lines and angles which go to determine the perimeters and areas of polygons and polyhedra, and many others.

§37

Many as the faults are, with which I believe we have not unjustifiably reproached past expositions of the *doctrine of the infinite*, it is nevertheless a matter of common knowledge that we *usually obtain quite correct results* if we follow the generally 65
accepted rules of infinitary calculation with suitable precautions. Such results could never have appeared had there not existed some unimpeachable mode of conceiving and conducting these calculations; and I am quite ready to believe that this unimpeachable mode was hovering in the minds of the sagacious discoverers of the method, and at bottom only this mode, even if it be true that they were not yet in a position to set forth their thoughts with all the desirable clearness: an aim which in difficult cases usually requires for its attainment a number of preliminary experiments.

May I be allowed, therefore, to sketch here in rough outline how I think it necessary to conceive this method of calculation, in order to justify it completely. It will suffice to speak of the process followed in the so-called *differential* and *integral calculus*, for the methods of calculation with the infinitely great follow at once by mere antithesis, particularly after all that *Cauchy* has achieved in the matter.

I have no need, then, of so restrictive a hypothesis in this matter as the one so often considered necessary, to wit: that the quantities to be calculated with can become *infinitely small*. By such a restriction we should be excluding at the outset all bounded temporal quantities from the scope of this method, all bounded spatial quantities, all the forces of bounded substances, in short, all the quantities whose determination matters most to us. I ask one thing only: that these quantities, when they are *variable*, and not independently variable but *dependently* upon one or more other quantities, should possess a *derivate*, or what Lagrange calls 'une fonction dérivée'—if not for every value of their *determining* variable, then at least for all the values to which the process is to be validly 66
applied. In other words: when x designates one of the independent variables, and $y=f(x)$ designates a variable dependent upon it, then, if our calculation is to give a correct result for all values of x between $x=a$ and $x=b$, the mode

of dependence of y upon x must be such that for all values of x between a and b the quotient

$$\frac{\Delta y}{\Delta x}=\frac{f(x+\Delta x)-f(x)}{\Delta x}$$

(which arises from the division of the increase in y by the increase in x) can be brought as close as we wish to some constant, or to some quantity $f'(x)$ depending solely upon x, by taking Δx sufficiently small; and subsequently, on our making Δx smaller still, either remains as close thereto or comes closer still.*

Once an equation between x and y is given it is usually a very easy and well-known matter to find this derivate of y. If for example

$$y^3=ax^2+a^3, \qquad (1)$$

then we should have for every Δx other than zero

$$(y+\Delta y)^3=a(x+\Delta x)^2+a^3, \qquad (2)$$

whence by the known rules

$$\frac{\Delta y}{\Delta x}=\frac{2ax+a\Delta x}{3y^2+3y\Delta y+\Delta y^2}=\frac{2ax}{3y^2}+\frac{3ay^2\Delta x-6axy\Delta y-2ax\Delta y^2}{9y^4+9y^3\Delta y+3y^2\Delta y^2},$$

and the *derived function* of y, or in Lagrange's notation y', would be discovered to be

67

$$\frac{2ax}{3y^2},$$

a function obtained from the expression for

$$\frac{\Delta y}{\Delta x}$$

by first suitably developing it, namely into a fraction whose numerator and denominator separate the terms multiplying Δx and Δy from the terms which do not, and then putting Δx and Δy equal to zero in the expression

$$\frac{2ax+a\Delta x}{3y^2+3y\Delta y+\Delta y^2}$$

thus arrived at.

I have no need to speak of the manifold advantages of finding this *derivate*, or of how by its means we can calculate the finite

* It can be shown that all *dependently variable quantities* which can be *determined* at all are subject to this law with the proviso that exceptional values, though they may occur to an infinite number, can only occur at *isolated values* of the *independent variable*. [But see page 54 of this volume.]

increment of y corresponding to a finite increment of x, or yet of how, when only the derivate $f'(x)$ is given, we can identify the primitive function $f(x)$ save for an [additive] constant.

Having now seen that the derived function of a dependent quantity y with respect to its variable x can be obtained by first developing

$$\frac{\Delta y}{\Delta x}$$

in such wise that neither Δx nor Δy appear as divisors, and then putting Δx and Δy both equal to zero, we shall not deem it so very inappropriate to symbolise the derivate by

$$\frac{dy}{dx},$$

provided we make two things clear: (i) that all the Δx and Δy which occur in the development of $\Delta x : \Delta y$ (or if you like, the dx and dy written in their stead) are to be regarded and treated as 68
mere zeros; and (ii) that the symbol $dy : dx$ shall not be regarded as the *quotient* of dy by dx, but expressly and exclusively as a *symbol* for the derivate of y with respect to x.

It is evident that such a process cannot be reproached with assuming ratios between inexistent quantities (of zero to zero) for the above symbol is not put forward for any other interpretation whatever than as a *mere symbol*.

It will be equally irreproachable to denote by

$$\frac{d^2y}{dx^2}$$

the *second* derived function of y with respect to x, meaning that quantity depending on x (or possibly constant) to which the quotient

$$\frac{\Delta^2 y}{\Delta x^2}$$

comes as close as we like, provided we can take Δx as small as we like; and to interpret this as follows: (i) that the Δx and $\Delta^2 y$ which occur in the development of $\Delta^2 y : \Delta x^2$ are to be regarded and treated as *mere zeros*; and (ii) that we must see in the symbol $d^2y : dx^2$ not a division of zero by zero, but only the *symbol* of the function obtained from the development of $\Delta^2 y : \Delta x^2$ by making the alteration just demanded.

Once these meanings of the symbols dy/dx and d^2y/dx^2 . . . have been laid down in advance, we can rigorously prove that every variable quantity which depends in a definable manner on another independent variable x,

69 $$y = f(x),$$

is compelled to satisfy the equation

$$f(x+\Delta x) = f(x) + \Delta x \frac{df(x)}{dx} + \frac{\Delta x^2}{1.2}\frac{d^2f(x)}{dx} + \frac{\Delta x^3}{1.2.3}\frac{d^3f(x)}{dx} + \ldots$$
$$\ldots + \frac{\Delta x^n}{1.2.3\ldots n}\frac{d^nf(x+\mu\Delta x)}{dx^n},$$

with $\mu < 1$, save at the very most for certain isolated values of x and Δx.*

Everybody knows how many important truths of the general theory of quantity, and above all of the so-called higher analysis, can be established by means of this one single equation. But this same equation also clears the way to the solution of the most difficult problems in the application of the theory of quantity—in the doctrine of space or geometry, in the doctrine of forces or statics and mechanics and so forth, in the rectification of curves, in the complanation of surfaces, in the cubature of solids—and all this without any supposition of the infinitesimal, which would in these cases be a contradiction, and without any other alleged principle such as the well-known one of Archimedes and so forth.

Now if it is legitimate to set up such equations as that for the rectification of a curve in rectangular coordinates:

$$\frac{ds}{dx} = \sqrt{\left[1 + \left(\frac{dy}{dx}\right)^2 + \left(\frac{dz}{dx}\right)^2\right]}$$

in the sense described above, then we shall incur no danger of error in setting up equations of the following kind:

70 $$d(a+bx+cx^2+dx^3+\ldots) = bdx + 2cxdx + dx^2dx + \ldots, \quad (sic)$$
$$ds^2 = dx^2 + dy^2 + dz^2;$$

or if r denotes the radius of curvature of a plane curve,

$$r = -\frac{ds^3}{d^2y \, . \, dx},$$

* A proof of this theorem for all modes of dependence of y upon x, whether capable or incapable of representation by symbols hitherto in use, has long since been written out by the author, and will perhaps be published in the near future. H. (Note by the posthumous editor, Dr. Přihonský.)

and so forth; whereby we regard the symbols *dx*, *dy*, *dz*, *ds*, d^2y and so forth never as the symbols of actual quantities, but always as equivalent to zero, and consider the entire equation to be nothing but a *compound symbol so constituted that* (i) *if we carry out only such changes as algebra allows with the symbols of actual quantities* (in this case, therefore, also divisions by *dx* and the like) *and* (ii) *if we finally succeed in getting rid of the symbols dx, dy and so forth on both sides of the equation, then no false result will ever be the outcome.*

This fact and its necessity are easily understood. For if the equation

$$\frac{ds}{dx}=\sqrt{\left[1+\left(\frac{dy}{dx}\right)^2\right]}$$

is irreproachable, for example, why not also the equation

$$ds^2=dx^2+dy^2,$$

since the former is derived from the latter by just that process?

No effort is needed, finally, to notice that we are safe from error in dealing with equations which contain the symbols *dx*, *dy*, . . . if we abbreviate the process and omit at the very outset all the addends of which we know for certain that they will disappear in the end as being equivalent to zero. So soon, for 71
example, as any sort of calculation has led us from (1) and (2) to the equation

$$3y^2\Delta y+3y\Delta y^2+\Delta y^3=2ax\Delta x+a\Delta x^2,$$

which the passage to zero-equivalent symbols transforms into

$$3y^2dy+3ydy^2+dy^3=2axdx+adx^2,$$

we perceive immediately that the addends containing higher powers dy^2, dy^3, dx^2 will in any event disappear at the finish, so that we may as well write

$$3y^2dy=2axdx$$

at once, and obtain at once the desired derivate

$$\frac{dy}{dx}=\frac{2ax}{3y^2}$$

with respect to *x*.

Let us close by putting the whole process in a nutshell: it rests on principles altogether similar to those behind our calculations with the so-called *imaginary quantities* (which like our *dx*, *dy*, . . . are mere symbols) or to those behind the recently invented short methods of division and other abridgements of calculation.

It is sufficient in all these cases, as in our own case above, to give notice that the symbols we introduce,

$$dx, \frac{dy}{dx}, \frac{d^2y}{dx^2}, \ldots, \sqrt{-1}, (\sqrt{-1})^3, \frac{\sqrt{-1}}{-\sqrt{-1}}, \ldots, \text{etc.},$$

shall be given only such a meaning, and shall be subject only to such changes, that when the non-objective symbols give place to others denoting quantities, the two sides of the equation always end by being really equal to one another.

72 §38

Turning now to the applied parts of the theory of quantity, the first paradoxes in the *doctrine of time* are encountered in the idea of *time itself*, particularly in so far as time is to be a *continuous extension*. But since the idea of *continuous extension* has been famous down the ages for *apparent contradictions* which are equally awkward for temporal, for spatial and even for material continua, we intend to deal with all three simultaneously.

It was recognised, to be sure, that everything extended had to be composed of parts, and recognised too that the existence of extended wholes could not be explained, without a vicious circle, by the composition of parts themselves already extended; but a contradiction was nevertheless alleged to lurk in the hypothesis that it arises out of parts devoid of extension and absolutely incomposite: instants in time, points in space and (in the realm of the actual) atoms or simple substances.

Enquiries into the objectionableness of this hypothesis were met in a variety of ways. At one time it was asserted that a property absent from each individual part could not accrue to the whole; at another time that every two instants or points or substances still stand at a certain distance apart and hence fail to constitute a *continuum*.

A very slight reflexion indeed suffices to discern the absurdity of these objections. What—is a property absent from each individual part not to accrue to the whole? Rather the exact opposite! Every whole has, and must have, properties lacking in
73 its separate parts. An automaton has the property of almost deceptively mimicking the movements of a living person, whereas its separate parts, its springs, its little wheels and the like, do not possess any such property.

As to the other objection, it must indeed be allowed that any two instants or points or (in the realm of the actual) any two substances are separated by an infinity of others; yet what precise contradiction follows from this? Only this: that nothing extended can be generated by a solitary pair, or a triad, or a tetrad, or any merely *finite* set of such entities. We ourselves concede all this and more still, namely, that even an infinity of points is not always sufficient for the generation of a continuum, to which end the points must also be properly *arranged*. If we try to form a clear idea of what we call a '*continuous extension*' or '*continuum*,' we are forced to declare that a continuum is present when, and only when, we have an aggregate of simple entities (instants or points or substances) so arranged that each individual member of the aggregate has, at each individual and sufficiently small distance from itself, at least one other member of the aggregate for a neighbour. When this does not obtain, when so much as a single point of the aggregate is not so thickly surrounded by neighbours as to have at least one at each individual and sufficiently small distance from it, then we call such a point *isolated*, and say for this reason that our aggregate does not form a continuum. If on the contrary a given aggregate contains no such point, so that taking any of its points and any 74 chosen distance, that point always has another member of the aggregate for its neighbour at that distance, no reason is left for our denying the name of a continuum to the given aggregate. For what else could we wish to demand?

'Why, this much,' they answer, '*that every point have a neighbour in immediate contact with itself*!' This requirement, however, is an obvious impossibility and harbours a contradiction. For when are you going to say that a pair of points touch each other? Perhaps when the boundary of the one, say its right limb, coincides with the boundary of the other, say its left limb? But points are the simple constituents of space and thus have no boundaries, no right or left limbs. Either the one point has a part in common with the other, and then it coincides altogether; or it is something different therefrom, and the two must lie altogether apart, leaving room for an intermediate point; room in fact for infinitely many intermediate points, since the argument can be applied to the first intermediate point over again.

'*This is all very incomprehensible*,' it is said. Admittedly, it cannot be grasped by the fingers, and cannot be perceived by the eye, but it can be cognised by the intellect, and for something which is what it is of necessity, and which could not be otherwise. A contradiction can thus be imported only when we represent it to ourselves as something other than what it is, when we represent it to ourselves falsely.

'How impossible it is to grasp,' people go on to say, 'the idea of a line as an accumulation of infinitely many points, and even as an infinite set of such accumulations, as we have to do according to the received doctrine! For even the shortest line is supposed divisible into an infinite set of other lines, namely by

75 iterated halvings.' In all this collection of thoughts I find nothing erroneous and nothing startling, save only for the expression 'shortest line,' an expression which many, no doubt, have allowed to escape them for want of attention, because after all there is and can be no such thing, and because the very line they speak of is supposed on this very occasion to be divisible into smaller ones. Every infinite set whatever, and not only the set of points on a line, can be partitioned into subsets which are themselves infinite sets, and even into infinitely many such subsets. For if ∞ denotes an infinite set, then

$$\frac{\infty}{2}, \frac{\infty}{4}, \frac{\infty}{8}, \ldots$$

are also *infinite sets.* This inheres in the idea of the infinite.

If the foregoing explanations have proved satisfactory upon a more prolonged consideration, one final question might still be posed: how are we to interpret the assertion of those mathematicians who declare both that extension can never be generated by the mere accumulation of points however numerous, and also that it can never be resolved into simple points? Strictly speaking, we should *on the one hand* certainly teach that extension is never produced by a finite set of points, and produced by an infinite set only when, but always when, the afore- and oft-mentioned condition is fulfilled—namely that each point in the set has, at each sufficiently small distance, a neighbour also belonging to the set; and *on the other hand* we should admit that not *every partition* of a spatial object works down to its simple parts: in no case one whose subsets are finite in number, and not even all that subdivide to infinity, say by successive halvings. Nevertheless, we must still insist that every continuum can be made up in the last analysis of points and of points alone. Once the two [apparent opposites] are properly understood, 76 they are perfectly consistent with one another.

§39

It was only to be expected that particular stumbling blocks would be found in that particular continuous extension which is called time. The doctrine of time was bound to provide with welcome matter those philosophers who, like the sceptics, lay themselves out to confuse human ideas instead of clarifying them, and to nose out contradictions in every nook and cranny. On the present occasion we shall mention only the most important points, and this the more, in that not everything brought forward in this connexion touches on the idea of the infinite.

The problem was raised whether time be anything *actual*, and if actual, then whether substance or accident, and if substance, whether created or uncreated.

'If created,' they argued, 'then it must have had a beginning, and will presumably come to an end, thus changing and requiring a super-time in which to change. It is still more absurd to hold that time be *identical with God*, or an accident present in God. It is quite true, moreover, that we make an antithesis *between time and eternity*; but what then is eternity? How is it possible that even the smallest stretch of time, even the twinkling of an eye, after which a "moment" is named [in German: "*Augenblick*"], should not only contain an infinite set of instants, but even an infinite set of *time-intervals*?' They even ended by saying that 'There is *no such thing as time*; for past time, just because it is past, no longer is; future time, just because it is future, is n o t yet; and what is present, is nothing but a *mere instant* in the strictest sense of the word, enjoying no *duration* and 77 no title to the *name* of time.'

According to my ideas, and in the strict sense of the word, time is admittedly *not an actual entity*. In the strict sense of the word we attribute actuality only to *substances* and their *forces*. Time, therefore, I hold to be neither God Himself, nor a created substance, nor an *accident* in God, nor an accident in any created substance, nor an accident in any aggregate of created substances. For this very reason, time is something *not subject to change*, but rather that within which all change takes place. If

the opposite seems to be said, as in the proverb that 'times change,' we have been long since reminded that by 'time' we here understand the things contained in time, together with their states and conditions. To come now to my own definition, I say that time is that *determination* of any and every variable (or what is synonymous therewith, dependent) substance, whose mental representation must be conjoined with that of the substance in order that, given two *contradictory attributes b* and non-*b*, we may truly predicate the one attribute of the substance in question and negate the other. More precisely, the *determination* here mentioned is a *single simple particle* of time, a moment or instant, at which we must place the substance *x* of which we desire to predicate with certitude just one of each pair of contradictory attributes *b* or non-*b*. Our assertion should therefore read more properly as follows: At the moment *t*, the substance *x* either possesses the attribute *b* or possesses the attribute non-*b*. Once the correctness of this definition of the idea of an instant is granted me, I can clearly state what time is, and indeed, what '*all time*' or *eternity* is: it is that particular whole to which all instants belong as members. All *finite* times, all *durations* or *time-intervals* contained between two given moments are defined by me as aggregates of all the moments lying between those two
78 given terminal moments. Conformably with these definitions there is no difference between eternity and time when the latter is conceived no longer in the usual manner as finite, but as bilaterally interminate. But there does exist a great difference between the manner in which God, and the manner in which variable or created beings, are situated with respect to time. The latter are immersed in time *as the locus of their variation*, whereas God is at all times changelessly the same. This has given occasion to call God alone *eternal*, and to call all other beings, all the creatures of God, *temporal beings*.

It may perhaps be a difficult task for *our imagination* to picture in sense-derived images how even the shortest duration, the mere twinkling of an eye, can already contain an infinite set of time-intervals. It is enough for our *intellect* to grasp it and recognise it as something that cannot be otherwise than it is. The objective foundation for this can also be discerned in the idea of time sketched above, but its exposition would digress too far for the present occasion. The only absurdity would lie in asserting that the short duration contained the same number (die gleiche Menge) of instants as the longer duration, or that the time-intervals into which a short duration were resolved had the same length as in the case of the longer duration.

In conclusion, the fallacy of the attempt altogether to destroy the idea of time lies so near the surface that it is hardly worth a word of refutation. We admit of course that time is in no sense an existent entity, and that existence accrues neither to time past nor to time future, not even the present having any; but how does it thence follow that time be a *non-entity*? Are not 'absolute propositions and truths' at least Somethings, even though nobody lets it enter his head to maintain that they are Existents?—unless indeed they are being confused with their intake into the consciousness of a thinker, and therefore with 79
thoughts or *judgements*.

§40

Turning to the paradoxes in the *doctrine of space*: it is well known that we have not succeeded in explaining space. It has often been held to be, like time, something existent; it has been confounded with the substances situated within it; it has even been identified with God, or at least regarded as an attribute of the divinity; even the great *Newton* lapsed into the thought of defining space as the divine sensorium; people have not only supposed that substances often *move* about space, but even that space itself moves, that is, places change places; since *Descartes*, we are believed to have discovered that not all substances, but only the so-called material substances are located in space; then *Kant* lit upon the unfortunate notion, still repeated after him by many, that both space and time have no objectivity, but are merely *subjective forms of our intuition*; more recently still, the question has been mooted whether other beings do not have another space, say with two or with four dimensions; and as the climax of it all, *Herbart* has enriched us with a double space, a *rigid and* [a] *continuous space*, together with a correspondingly double time. My opinions on all these subjects have already been stated elsewhere.

Space, and time like it, is in my view *not a quality* (Beschaffenheit) of substances, but a *determination* of them only, namely as follows: I give the name 'places occupied' to those determinations of created substances which give the grounds why, being in

possession of such and such qualities at a given moment, they
80 produce such and such changes in one another; and then define space, the whole of space, as the aggregate of all places occupied. This definition made it possible for me objectively to deduce the teachings of the science of space from those of the science of time, and thus to show, for example, the fact, and the reason for the fact, of space having three dimensions, and other things besides.

The paradoxes, therefore, which people have found in the very *idea* of space, in that *objectivity* which it is supposed to possess in spite of its not being anything actual, in the infinite set of its particles, and in the continuous whole which they form when taken together in spite of no two particles being in immediate contact—are just apparent contradictions which I do not consider myself obliged to pass in review, but rather entitled to regard as settled.

The first thing to demand closer consideration is perhaps the idea of the *magnitude* of a spatial extension. All are agreed that all extension possesses *magnitude*, and also that both in the single temporal dimension and in the threefold spatial dimension the quantities encountered can only be determined by their ratio to some one quantity arbitrarily chosen as the unit of measurement, and furthermore, that this chosen unit must be homogeneous with the mensurand—therefore be, for lines a line, for surfaces a surface, for bodies a body.* But if we go on to ask wherein the so-called magnitude of a spatial extension rightly consists, we are tempted to conclude that by magnitude we must mean simply the [numerosity of the] set of points of which it consists—and this all the more, seeing that it does consist of nothing but points, and that quantity never attends to the arrangement of the members of an aggregate, but exclusively to the set of those members (14). The *very name* we give to the magnitude

* Not a few readers would probably like to see a definition of these three kinds of spatial extension. Supposing that the definition given in §38 of *extension as such* has been accepted (and it has the merit of requiring but little modification to make it applicable to what the *general theory of quantity* calls *continuous variables*) I say as follows. A spatial extension is *simply* extended (or
81 a line) if each point has at (13) each sufficiently small distance one or more neighbours, yet not so many that the neighbours at this distance form an extension by themselves; it is *doubly* extended (or a surface) if each point has at each sufficiently small distance a whole line-ful of neighbours; and it is *triply* extended (or a *volume*) if each point has at each sufficiently small distance a whole surface-ful of neighbours.

of a surface or volume seems to strengthen the temptation, for we call it their '*content*.' Nevertheless, a closer study reveals that this is not true. How could we, if it were true, so generally and so unhesitatingly suppose that the volume (say) of a cube is not affected by our including or excluding the boundary surfaces, which after all have a magnitude of their own? We do this beyond a doubt when we say that a cube of side 2 has eight times the volume of a cube of side 1, in spite of the circumstance that [when eight such smaller cubes are assembled to form a single large one] the abutment in pairs of the twenty-four interior squares causes the [partitioned] large cube to possess twelve fewer units of boundary surface area than do the eight smaller ones [taken disjunctively].

We see from this that by the magnitude of a spatial extension, whether linear or superficial or solid, we mean nothing else than 82
a quantity so derived from a unit homogeneous with the mensurand that if the quantity m arises on the chosen scheme from the piece M, and the quantity n from the piece N, then precisely the quantity $m+n$ arises on that same scheme from the piece formed by juxtaposing M and N, irrespectively of whether the boundaries of M and N and $M+N$ are counted in or not. The treatise mentioned in §37 shows that the most general formulae which the science of space works out for rectifications, quadratures and cubatures can be derived from the above definition without appealing to any other hypothesis, and in particular without appealing to the falsely-so-called principles of Archimedes.

§41

Taking our stand on the explanations so far given, we can put forward the following statements—for all their paradoxical contrariety to ordinary ways of thought—with no fear of being convicted of self-contradiction.

1. The aggregate of all points lying between two points a and b represents a one-dimensional extension or a line. The length of this straight line is exactly the same, whether we include both points a and b and so make our line a *bounded* one, or whether we exclude one or both of those points and so make our line an *unbounded* one. Every such unbounded straight line has, in the direction where a terminal point is lacking, for this precise reason no *outermost* or farthest point; instead, there always lies

83 another point behind any one point—yet always at a finite distance.

2. The perimeter of the triangle *abc* can be dissected into (i) the bilaterally terminated straight line *ab*, (ii) the straight line *ac* terminated at *c* but not at *a* and (iii) the straight line *bc* terminated neither at *b* nor at *c*. Yet this perimeter equals the sum of the three lengths *ab*, *bc* and *ca*.

3. Let the straight line *az* be bisected at *b*, the piece *bz* bisected at *c*, the piece *cz* bisected at *d* and so on without end. Let a first aggregate be formed of all the points remaining *after* the removal of *z* itself together with the infinitely many points of bisection *b*, *c*, *d*, . . . and so forth. This first aggregate still merits the name of a *line*, and its magnitude will be the same as before. But if we form a second aggregate by restoring *z*, the new whole is no longer a continuous extension; for of no distance however small can we say that, at *that* distance, and at every *smaller* distance, *z* has some member of this second aggregate for a neighbour. The reason is, that at all distances of the type $az/2^n$ a neighbour is wanting, so that *z* turns out to be an 'isolated point.'

4. If the distance between the points *a* and *b* equals that between α and β, then the set of points between *a* and *b* must be put *equal* to the set of points between α and β.

5. Extended figures having equal sets of points are also equal in magnitude. The converse, that extensions equal in magnitude have equal sets of points, is not true.

6. When a pair of spatial objects are completely similar [in the geometrical sense], their two sets of points must have the same ratio as their two magnitudes.

84 7. If the ratio of the magnitudes of two perfectly similar spatial objects is irrational, then the ratio between their sets of points is also irrational. *Sets* exist, therefore, *whose ratio is irrational in any of the possible manners*; but such sets will always be infinite ones.

§42

So far as I know, only the sixth of these propositions (which could obviously be multiplied without difficulty) has received notice in the literature of mathematics, and this only by the putting forward of the contrary proposition that *similar lines, howsoever great their difference in magnitude, still contain an equal set of points*. This was asserted in particular of similar and concentric circular arcs by Dr. *J. K. Fischer*, in his 'Outline of the Whole of Higher Mathematics,' (15) Leipzig 1809, volume II, note to §51. The reason he gave was, that through each point in the one a radius can be drawn to meet a corresponding point in the other. *Aristotle*, as is well known, already busied himself with this paradox. Fischer's reasoning betrays the opinion that two infinite sets, as well as two finite ones, are bound to be equal so soon as each member of the one can be coupled off against one member of the other. The disclosure of this mistake makes it unnecessary further to refute Fischer's opinion, of which it can be said in addition that, were it true at all, there would be no point in restricting it to circular arcs that are similar and concentric, or to circular arcs at all, since the same grounds would hold good for all *straight lines* as well as for a great diversity of by no means similar curves.

§43 85

Scarcely any truth in the science of space can have been sinned against so often by the teachers of that subject, as the truth that *every distance between two points in space, and therefore every bilaterally bounded straight line, is only finite*, or in other words: stands to every other such distance in a ratio which is exactly determinable in terms of pure concepts. Indeed, hardly a geometer exists who has not spoken at some time or other of *infinitely great distances*, or not asked that in some circumstances or other a straight line, though supposed to be terminated in both directions, should become *infinitely great*. As for examples, let us content ourselves at present with a reference to the pair of lines which are called, in the geometrical sense of the terms, the 'tangent' and the 'secant' of an arc or angle. According to the express definition, these are supposed to be a pair of straight lines limited in either direction; and how few there are who scruple to teach that the tangent and the secant of a right angle be *infinitely great*? Yet the nemesis of this false doctrine overtakes us at once, in the shape of our embarrassment when faced with the question whether these two infinitely great quantities are to be regarded as *positive* or *negative*. For of course, any reason in favour of the one decision would also be in favour of the other, because, as is well known, a parallel through the centre of a circle to one of its tangents is related to that tangent in a perfectly symmetrical fashion, and can therefore no more cut in on the one side than on the other. Nor does the quantitative expression for these two lines,

$$\frac{1}{0},$$

afford any grounds for declaring them positive or negative, since zero itself is neither positive nor negative. To suppose the
86 existence of an infinitely large tangent for the right angle, or for all angles of the type

$$\pm n\pi \mp \frac{\pi}{2},$$

is consequently not merely paradoxical, but out and out false (16).

Be it noted in passing that the angles 0 and $\pm n\pi$ also fail to possess a tangent, or a sine. The difference between the two cases is simply this: that in the former, no error can ensue from the sheer ignoration of products in which these quantitative expressions appear as factors; whereas in the latter, where they appear as divisors, we conclude that the formulae are making an unlawful demand.

§44

Equally unjustified was the process, imitated fortunately by very few people, by which *Johann Schulz attempted to compute the size of the whole of infinite space.* From the two premises (i) that from any arbitrary point *a* as vertex we can draw straight lines to infinity in every conceivable direction, and (ii) that every point *m* we can think of in the whole of universal space must lie on one and only one of these lines, he thought himself entitled to infer that the whole of infinite space could be regarded as a *sphere* with *a* for centre and ∞ for radius, which in his eyes entailed at once that the whole of infinite space must have exactly the volume

$$\frac{4}{3}\pi\infty^3.$$

87 Now if this could be justified, it would be one of the most important theorems of the science of space. Nor can any really solid objection be raised against the two premisses; of which I must confess that I have not reproduced them verbatim from *Schulz*' own exposition, this not being available to me. Were

anyone, in fact, to impugn the second premiss on no other ground than that it entails a very unequable distribution of points in space—to wit, by their much denser concentration around what is after all the arbitrary point *a*—he would only betray that he had not yet overcome the prejudice attacked in §21 above. *Schulz* fell into error in one point only (but then completely) and that point was his assumption that the straight line drawn from *a* as vertex to infinity in all directions, and passing eventually through each point in space, were in fact *radii*, and hence bilaterally limited. This was, after all, the hypothesis from which the spherical form of infinite space was deduced, and the calculation of its volume as

$$\frac{3}{4}\pi\infty^{3} \qquad (sic)$$

was carried out. This mistake gives rise to yet other absurdities; because every sphere admits to be sure of a circumscribing cylinder, a circumscribing cube, as well as many other spatial entities such as infinitely many tangent spheres of equal diameter, so that the alleged 'whole' of space would not be the whole at all, but only a part having infinitely many other spaces outside itself.

The futility of most of the paradoxes (*mysteria infiniti*) put forward by *Boscovich* (17) in his '*Dissertatio de transformatione locorum geometricorum*' (published at *Rome* in 1754 as an appendix to
Volume III of his '*Elementa universae Matheseos*') is adequately 88
revealed by the one single remark that a line drawn to infinity, even if only in one direction, is for this very reason not bounded in the direction concerned, and that we can therefore no more speak of a terminal point for it than we can speak of the vertex of a sphere, of the curvature of a straight line or an individual point, or of the intersection of two parallels.

§45

Infinitely small distances have been assumed hardly less often than infinitely large ones, especially when it seemed necessary to treat as straight (or plane) those lines (or surfaces) of which no portion was both extended and really straight (or plane) on the plea for example of more easily determining their arcual length or the magnitude of their curvature, or other features of theirs noteworthy from the point of view of mechanics. People even went so far as to invent fictitious distances supposed to be measured by infinitely small quantities of the second order, of the third order, or even of still higher orders.

Now if this process seldom led to false results, particularly in geometry, the only circumstance we can thank for it was the one already mentioned in §37, that in order to apply to *determinable* spatial extensions, variable quantities must be so constituted that, with the exception at most of *isolated individual values*, their first and second and all subsequent *derivates* exist. For if they do exist, then what was being asserted of the 'infinitely small' lines and surfaces and volumes can, as a general rule, quite rightly be asserted of all lines and surfaces and volumes which, while always remaining finite, nevertheless can be taken as small as we please, or as we express it, can be infinitely decreased. The former assertions, mistakenly applied to 'infinitely small distances,' were thus really true of the 'variable quantities.'

It can be readily understood, however, that such methods of describing the situation were bound to produce, and appear to
89 prove, much that was paradoxical and even quite false. How scandalous it sounded, for example, when they said that every curve and surface was nothing but an assemblage of infinitely many straight lines and plane surfaces, which only needed to be considered present in infinite multitude; and aggravated even this by adding the supposition of infinitely small lines and surfaces which were themselves curved. How strange it was for them to say of a curve which had no curvature at all in one of its points, e.g. in a point of inflexion, that its curvature at this point is infinitely small, and its radius of curvature therefore infinitely great; or for them to describe a cusp as a point where the curvature is infinitely great and the radius of curvature infinitely small, and many similar things besides.

§46

I only beg leave to quote one example of the absurdities for which the hypothesis of infinitely small distances provided both the matter and the opportunity. It shall be the very simple and striking example which, as *Kästner* tells us (in the Preface to Volume II of his (18) 'Elements of higher analysis'), was set forth by *Galilei* in his '*Discorsi e dimostrazioni matematiche,*' no doubt solely for the purpose of stimulating reflexion, to the effect that the *circumference of a circle is equal in magnitude to its centre.*

To obtain an idea of the attempted proof, let the reader imagine a square *abcd*, and a quadrant *bd* with *a* as centre and 90
ab as radius, together with a straight line *pr* parallel to *ab* cutting *ad* in *p*, the diagonal *ac* in *n*, the quadrant in *m*, and *bc* in *r*; in short, the usual figure for proving that the area of a circle with radius *pn* is equal to the area of the annulus left between circles with the radii *pm* and *pr*, or:

$$\pi \,.\, pn^2 = \pi \,.\, pr^2 - \pi \,.\, pm^2.$$

The nearer *pr* approaches to *ab*, the smaller becomes the circle with the radius *pn*, and the slenderer the annulus between those with radii *pm* and *pr*. Such geometers as did not boggle at infinitesimal distances transferred this situation to the case where *pr* comes infinitely near to *ab*, so that the distance *ap* becomes equal to '*dx*' and the equation

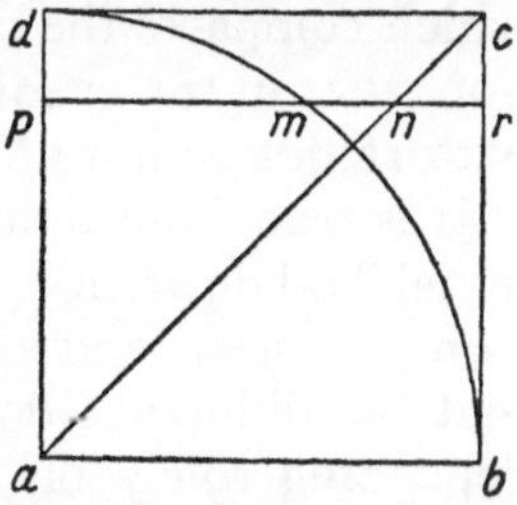

$$\pi \,.\, dx^2 = \pi \,.\, a^2 - \pi(a^2 - dx^2)$$

is supposed to hold, and does hold as being an identity. In this case, however, they imagined the circle with radius *pn* to have become infinitely small to the second order, whereas the ring between those with radii *pm* and *pr* now had a width only equal to

$$-mr = \frac{1}{2}\frac{dx^2}{a} - \frac{1}{2.4}\frac{dx^4}{a^3} - \ldots,$$

which is itself infinitely small to the second order. When they went to the length of letting *pr* pass over completely into *ab*, the infinitely small circle with radius *pn* shrank to the point *a*, and the infinitely slender ring of width *mr* reduced to the bare peri- 91
phery of the circle with radius *ab*. Whence they seemed justified in inferring that the solitary centre-point *a* of any arbitrary circle with the radius *ab* had the same magnitude as the entire circumference.

The deceptiveness of this reasoning was chiefly caused by the intrusion of the infinitely small. The reader was thereby enticed into a series of thoughts which made it easier for him not to notice the absurdity of saying that, when our attention is transfered from *p* to *a* and no such radius as *pn* exists any longer, the point *a* still *survives as the midpoint* of the ultimate circle *pn*. Or again, not to notice the absurdity of saying that the ring left between the circle with the smaller radius *pm* and that with the larger radius *pr* 'ultimately' becomes the circumference of the previously greater circle, 'ultimately' meaning: when the two radii and hence the two circles have become equal. We are accustomed, to be sure, when dealing with infinitely small quantities, sometimes to regard two of them as equal, sometimes to regard the same two as differing by an infinitely small quantity of higher order and sometimes again as differing by zero. If we desire to argue according to the rules of logic, then we may only infer from the correctly written equation

$$\pi \, . \, pn^2 = \pi \, . \, pr^2 - \pi \, . \, pm^2,$$

which compares the mere areal contents of the circles in question, that in the event of *pr* and *pm* becoming equal, the circle with radius *pn* has no magnitude at all, and so ceases to exist.

It is true, indeed, and it follows from premises which I have myself laid down in §41, that we have circles which include their own periphery, and also circles which exclude it; and also true, that the difference does not affect their magnitude, the latter
92 depending solely on the radius. This may lead someone to attempt a fresh proof of Galilei's proposition, starting from the admittedly lawful demand that the circle with radius *pm* should exclude its periphery, and that with the radius *pr* include its periphery. For if this were done, the removal of the circle with radius *pm* from that with radius *pr* upon passing from *pr* to *ab* really would leave only the circumference of the quadrant. Even so, however, it would be impossible to speak of any circle centered at *a* and shrinking to a single point; and still less would it be permissible to appeal to the above equation for a proof that the point *a* and the said circumference were equal: because the equation in question only deals with the magnitudes of the three circles, irrespectively of whether their peripheries are counted in or not.

§47

I have already mentioned that the previous example was not put forward by its inventor as a truth to be marvelled at. What

they do teach as a serious theorem is, that the common cycloid has an infinitely great curvature (or synonymously, an infinitely small radius of curvature) at the point where it reaches down to its base line, and that it stands at right angles to that line. This is true enough in the sense that the radius of curvature decreases infinitely as the cycloidal arc approaches its base with infinite closeness, and that its direction is perpendicular at the actual point of attainment. What was added about infinitely small or vanishing radius of curvature amounts, when more rigorously expressed, only to this: that the curve, which is well known to proceed to infinity over its base in both directions and thus to possess *no boundary points*, here has two portions of arc coming 93
together in such wise that, both being perpendicular to the base, they form a *cusp* at which their directions are identical and form (as is not quite correctly said) a zero *angle* with one another.

At the same time it is quite possible to be persuaded on the strength of our calculations that such be the position of affairs, and yet not understand how it comes about, or even how it be at all possible. To make this clear as well, and only thus can we resolve the paradox, we must first grasp the reason why the common cycloid rises up perpendicularly from its base line.

This cycloid, then, is constructed as follows: at any point *o* on the base draw a circle equal to the generating circle and tangent to the base at that point, and marking off along it an arc *om* whose length equals the distance from *o* to the origin *a*, obtain *m* as one point on the cycloid. It follows at once that the angle *mao* approaches the more nearly to a right angle, the more nearly the point *o* approaches *a*, because the angle, *moa* is measured by the half of the arc *om* and thus continually decreases, while in the triangle *moa* the sides *oa* and *om* continually approach a ratio of equality. The angles adjacent to the third side *am* hence differ less and less from a right angle. The actual calculation shows this very clearly.

As further consequences we also have these: that the cycloidal arc *am* lies wholly on one side of the chord *am*, the side between it and the perpendicular *at* erected at *a*, and that this 94
perpendicular must indicate the direction of the curve at *a*. Describe next a circular arc with centre at *o* and radius *oa*. It will obviously cut the chord *om* externally, say at *r*, because we must have $or = oa > om$. Now let μ be a point on the curve lying

still nearer to *a*. In *ao* and nearer to *a* than *o* there will exist a point ω such that what we just said of the chord *om* can be repeated of the chord $\omega\mu$, to wit: the circular arc with ω as centre and ωa as radius cuts $\omega\mu$ externally, say at ϱ. As a result of $\omega a < oa$, the circular arc $a\varrho$ lies within the circular arc *ar*, and therefore between the cycloidal arc $a\mu$ and the circular arc *ar*. From thence we see that, howsoever small be the radius *oa* of a circular arc *ar* touching the cycloid *am* at *a*, there always exists another circular arc $a\varrho$ approaching still nearer to it in the vicinity. In other words, no circle however small could measure the curvature at *a*, did such a curvature exist at all. The truth of the matter is, therefore, that there is no curvature here whatever, and that the curve (which does not come to an end at this point) here possesses, as we already know, a cusp.

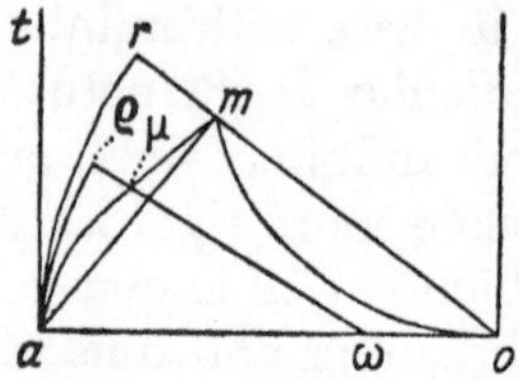

§48

Three other things have often been deemed paradoxical, namely: (i) many spatial extensions *spread themselves into infinite space* and contain points whose distance exceeds any one preselected distance, yet remain finite in magnitude; (ii) others are *confined within an altogether finite region* and contain no two points distant by more than a previously fixed amount, yet have an
95 *infinite magnitude*; and finally (iii) many spatial extensions retain a finite magnitude in spite of *winding infinitely often round a fixed point*.

1. As to the first point, we must first of all distinguish whether the spatial extension in question be a whole consisting of *separated parts* (like a hyperbola provided with four [*sic*] branches) or whether it be a *completely connected whole*, meaning: an extended whole, every extended part of which contains at least *one point* whose conjunction with the remaining parts would give rise to another spatial extension.

Nobody will see anything out of the way when an extension consisting of separated parts spreads itself into infinite space without having infinite magnitude; for he need only remember that a decreasing geometrical sequence of magnitudes also yields but a finite sum. In this sense, of course, even a line can remain finite while distributing itself to infinity—for example, if we mark off from a fixed origin *a* in a fixed direction *aR* a rectilinear segment *ab* and then, at intervals of constant length, mark off an infinite sequence of rectilinear segments each half as long as the preceding one, beginning with *cd*, the half of *ab*.

Now and henceforward, however, we shall speak only of spatial extensions which present themselves as *connected wholes.* This being understood, it is clear that among the *extensions of lowest category* (Art), among the *lines*, none will stretch to infinity without possessing at the same time an *infinite magnitude* (which will be a length). For this necessarily follows from the well-known truth that the shortest completely connected line linking 96
two given points is the straight line between them, and none other.*

The situation is different with *surfaces*, which despite constant

* The proof of this truth is so brief that I take leave to incorporate it into this note. If the line *amonb* is not straight, then somewhere upon it there must be a point *o* external to the straight line *ab*, and if we drop the perpendicular *oω* from *o* onto *ab*, certain distances must be unequal as follows:

$$a\omega < ao, \quad b\omega < bo.$$

Now since every point-pair forms a figure similar to every other point-pair, the points *a* and ω are joined by a curve *a*μω similar to the portion *amo* of the original curve *amonb*, and the points ω and *b* likewise by a curve ω*νb* similar

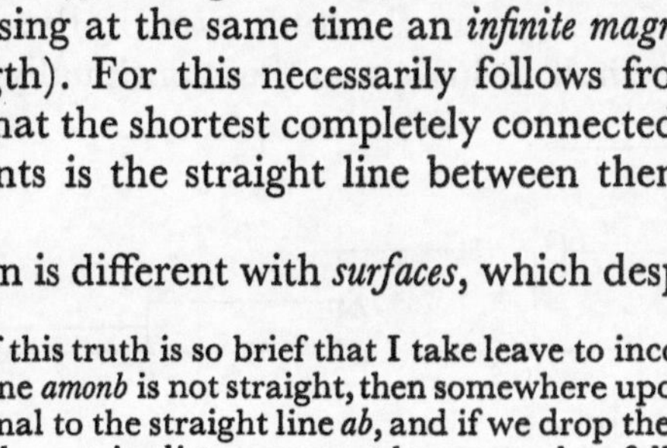

to the portion *onb* of the original curve. In virtue of this similarity, however, the length of the chord *a*ω bears to that of the curve *a*μω the same ratio as does the length of the chord *ao* to that of the curve *amo*, and the length of the chord ω*b* bears to that of the curve ω*νb* the same ratio as does the length of the chord *ob* to that of the curve *onb*. Furthermore, $a\omega < ao$ now entails $a\mu\omega < amo$, and $b\omega < bo$ now entails $b\nu\omega < bno$. The whole length *a*μω*νb* is accordingly less than the whole length *amonb*. The curve *amonb* is consequently not the shortest between *a* and *b*, because *a*μω*νb* is still shorter.

length can be made arbitrarily small in area by diminishing their breadth; as also with *spatial regions*, which despite constant length and breadth can be made arbitrarily small in volume by diminishing their height. This makes it intelligible why surfaces of infinite *length* and volumes of infinite *length and breadth* can
97 sometimes claim only a finite magnitude. We shall provide even the least expert reader with an example, if we ask him to take a straight line proceeding from the point *a* in the direction *aR* to infinity, and to mark off upon it the equal segments of unit length: $ab = bc = cd =$ and so forth without end; let him then erect on the first segment the square $b\alpha$, on the second the rectangle $c\beta$ only half as high, and thus continually adjoining rec-

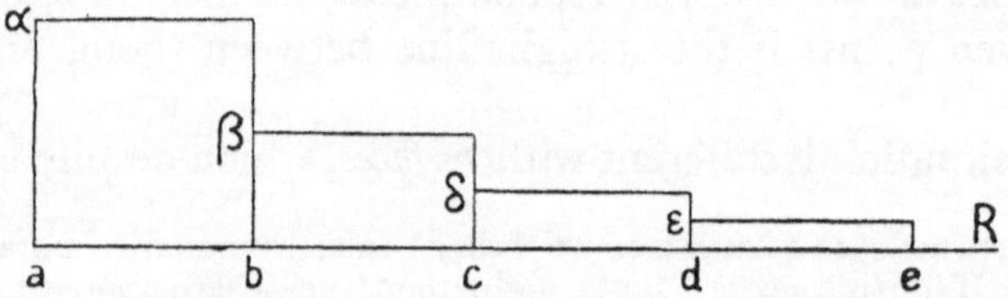

tangles each half as high as its predecessor. He will not be long in recognising that the connected surface now taking shape before his eyes stretches to infinity and nevertheless does not exceed 2 in area. Equally easy to imagine is the example of a cube of unit side standing on a sequence of square plinths, each plinth having twice the edge and four times the base area of its predecessor, but only one eighth of its height. The length and breadth of this infinite sequence of supposited cuboids both increase to infinity, but for all that the volumes are successively
98 halved, and despite its infinite basis, the pyramid-like whole does not exceed 2 in volume.

2. The second of our three cases shows a certain contrariety to the first. This first case, of an extension finite in magnitude though infinite in one or two of its dimensions, can occur with *surfaces* and *spatial regions*, but never with curves; whereas the second case, of an extension infinite in magnitude though enclosed in a finite space, can occur with *curves* and *surfaces*, but never with spatial regions. A spatial region no two of whose points are more distant than a certain fixed amount cannot have an infinite volume. Such is the conclusion to be drawn at once from the well-known fact that the greatest solid figure, no two

of whose points have a mutual distance exceeding E, is a sphere with E for diameter; for such a sphere contains all the possible points in question and has the volume $\pi E^3/6$, whereas every other solid which is confined within the same bounds must have a volume *less* than $\pi E^3/6$.

In contrast with this, an arbitrarily small surface such as a *square foot* can contain infinitely many *lines*, each a foot long if we so desire, and by joining their ends we obtain a composite and connected line whose length is certainly infinite. Exactly the same can be said of an arbitrarily small *volume* such as a *cubic foot*, to wit: we can draw infinitely many *surfaces* within it, each a foot square if we so desire, and by joining their edges we 99
obtain a composite surface whose area will indisputably be infinite. Only those can be astonished at all this who forget that we do not measure lengths and areas and volumes with the same unit, or forget that even if an arbitrarily small line does contain infinitely many points, we must certainly assume that the set of points in a surface is infinitely many times greater than that in a line, and just as certainly assume that the set of points in a solid is infinitely many times greater than that in a surface.

3. The third paradox at the beginning of this paragraph was that of extensions which wind infinitely often round a fixed point and still retain a finite magnitude. We have just seen in the first of our three cases that if an extension of this kind is to have *one dimension* it must lie wholly within a bounded region of space. Once that condition is satisfied, however, there is nothing mysterious in the phenomenon of a finite length despite infinitely many windings, if only the windings begin at a finite size and decrease in the proper fashion to infinity. This latter requirement is made attainable in its turn by the fact that the windings are to be round a *mere point*, for in this way the distances of points on the whorls from the centre and from one another can shrink indefinitely—the circle itself then teaching us that the length of the whorls can also be decreased to infinity. The *logarithmic spiral*, or rather an arc of it proceeding from a fixed point towards the pole without ever falling into the pole, 100
is an example of this phenomenon which the reader can hardly have failed to think of for himself.

The restriction to a finite region of space can actually be

dropped so soon as a *surface* or a *volume* [rather than a line] is to wind infinitely often round a fixed point. A simple example can now be described quickly. Let the reader look upon the above spiral as a sort of axis of abscissae, and imagine an ordinate erected at each point, perpendicular to the plane of the spiral. The set of all these ordinates will constitute a surface of cylindrical type, which in the outward direction recedes to infinity, but in the inward direction winds infinitely often nearer and nearer to the centre without ever reaching it. The size of this surface will depend upon the law according to which we let the ordinates increase or decrease. If only the ordinates to the inward branch of the axis of abscissae are prevented from increasing to infinity, the corresponding portion of our surface will always remain finite—since surfaces are always finite when neither abscissa nor ordinates increase to infinity. Even the portion over the expanding branch will remain finite in area if only the ordinates decrease more swiftly than the arcual abscissae increase. To make our choice definite, let the spiral acting as axis of abscissae be a *natural* (19) one, so that the centripetal branch, measured from the end of the unit radius vector, will have the length $\sqrt{2}$. The quasi-cylindrical surface whose one edge is the spiral shall take for its other edge an arc adapted from the plane hyperbola of higher order with the equation $yx^2=a^3$. Before adapting it, we note that the area between this hyperbola and the x-axis amounts only to a^2 for the portion where $x \geqq a$, but increases to infinity with x decreasing to zero.
101 Choose then some particular $a>\sqrt{2}$, place the point $x=a, y=0$ against the end of the unit radius vector and wrap the x-axis along the spiral arc in such a direction that the rising hyperbola stands over the centripetal, the falling hyperbola over the expanding branch of the basic spiral. The former point $x=a-\sqrt{2}, y=0$ then falls upon the pole, and to it there now belongs a merely finite ordinate. The quasi-cylindrical area over the centripetal branch is now that from $x=a-\sqrt{2}$ forward to $x=a$ and therefore equal to

$$a^3\left(\frac{1}{a-\sqrt{2}}-\frac{1}{a}\right)=\frac{a^3}{a-\sqrt{2}}-a^2,$$

and the area over the expanding branch is now that from $x=a$ forward to infinity and therefore equal to a^2, so that by addition and due regard to sign the whole quasi-cylindrical area rising from the whole spiral amounts to

$$\frac{a^3}{a-\sqrt{2}},$$

and when a has the particular value 2, this whole surface has an area amounting only to $4(2+\sqrt{2})$.

The situation is very similar in the case of solid extensions. Care is needed, however, on one fresh point: if we allowed the breadth and thickness of a corresponding solid to increase in the centripetal direction of the basic spiral, it would come to interpenetrate both neighbouring whorls. To avoid this, and to have an example free from self-impenetration, we can pass over the other possibilities and simply start from the above-mentioned spiral-based quasi-cylinder, giving it everywhere a certain thickness, yet nowhere more than half the distance to the next whorl on the right or left.

§49 102

When figures extended in space possess infinite magnitude, their behaviour in this respect is so diverse and often so paradoxical that we must pay particular attention to at least a few among them.

What has already been said can be regarded as shedding light enough on such facts as: that a spatial object comprising an infinite set of points does not for that reason need to be a continuum; that the determination of the magnitude of a continuous extension is not the determination of the set of points it contains; that of two extensions considered equal in magnitude, one can still contain infinitely more or infinitely fewer points than the other; that one surface or solid can contain infinitely more or infinitely fewer lines or surfaces than does another surface or solid considered equal to the former in magnitude.

1. The first point to which we would direct the reader's attention is this: the set of points in an arbitrarily short straight line az must be deemed *infinitely greater* than the infinite set of points obtained by taking out of it first the terminal point a, then b at a suitable distance towards z, then c at a shorter distance from b, and so forth in such wise that the sum of the infinitely many distances ab, bc, cd, . . . comes to be less than or equal to az. For each of the infinitely many pieces ab, bc, cd, . . .

into which az is partitioned is itself a finite straight line, and can be handled just as az has been handled, that is to say: in each of these pieces another such infinite set of points can be designated, all of which belong simultaneously to az itself. Such an infinite
103 set of points must consequently be contained in the whole line az infinitely often.

2. Whenever two straight lines are *equal*, and indeed whenever two spatial extensions are similar and equal (that is, agreeing in every *attribute* which can be *conceptually expressed* by comparison with a fixed given distance) they must be credited with equal sets of points, provided that the *boundaries* are reckoned alike in both cases, corresponding boundary points being either both included or both excluded. The contrary would, in fact, only be possible if distances existed which were metrically equal yet contained between their terminals an unequal set of intermediate points. Such distances do not exist, for they would contradict the idea we attach to *geometrical equality*; the distance ac being called unequal to ab, and specifically '*greater*' than ab, only if b and c lie on the same side of a and b between a and c. In this case, all the points between a and b are points between a and c, but not conversely, since not all points between a and c are points between a and b.

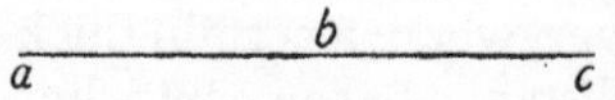

3. Let E denote the set of points lying between a and b, both included, and let the straight line ac have the integer n for its length measured by ab as unit. The set of points in the straight line ac, both terminals included, will then be equal to $n\,E-(n-1)$.

4. The set of points in the *surface of a square* of side 1 (the usual unit of area) including the periphery will be equal to E^2.

104 5. The set of points in the surface of a *rectangle* with one side m and the other side n and including the periphery will be

$$mnE^2-[n(m-1)+m(n-1)]E+(m-1)(n-1).$$

6. The set of points in the volume of a *cube* of side 1 (the usual unit of volume) including the periphery will be E^3.

7. The set of points in the volume of a *parallelepiped* of sides m, n, r and including the periphery will be

$$\begin{aligned}mnrE^3&-[nr(m-1)+mr(n-1)+mn(r-1)]E^2\\&+[m(n-1)(r-1)+n(m-1)(r-1)+r(m-1)(n-1)]E\\&-(m-1)(n-1)(r-1).\end{aligned}$$

8. We must attribute to a bilaterally unterminated straight line an infinite length and a set of points infinitely many times as great as the set of points in the unit straight line *E*. Moreover, an equal length and an *equal* set of points must be allotted to *every* such line: because, two such lines being determined by two equidistant point pairs one on each, those specifying point pairs constitute, not merely two similar figures, but also two figures geometrically equal.

9. Every particular point on such a line is situated in an exactly similar manner with regard to the parts of the line on either side of itself, and its situation offers no other conceptually expressible features than does the situation of any other point on the line. Yet for all that, we are not entitled to say that the point partitions the line into two portions *equally long*; for if we could say this of any point *a*, we could say it of any other point *b* for the

same reason, and so come to the contradiction that *aR* and *aS* were equal, but *bR* and *bS* unequal—the latter following from (20) *bR* equal to *aR* plus *ab*, together with *bS* equal to *aS* minus *ab*. We ought rather to make the assertion that a bilaterally unterminated straight line simply has *no midpoint at all*—no point, 105
that is, determinable solely by its conceptually expressible relation to this line.

10. We must attribute to the plane surface lying between two bilaterally unterminated parallel straight lines (that is, to the aggregate of all points contained in perpendiculars dropped from one parallel to the other) an *infinitely great* area and a set of points infinitely many times as great as the set of points in the unit area *E*. Moreover, an equal area and an equal set of points must be allotted to all such parallel strips possessing the same *width* (or length of inter-perpendicular): because here too, we can have determining elements which are not merely similar, but also geometrically equal—for example, two isosceles right-angled triangles equal in side and hypotenuse such that in each case one parallel is to lie along a side and the other to pass a vertex.

11. Every particular inter-perpendicular on such a strip is situated in an exactly similar manner with regard to the parts

of surface on either side of itself, and its situation offers no other conceptually expressible features than does the situation of any other inter-perpendicular. Yet for all that, we are not entitled to say that the inter-perpendicular partitions the surface into two *geometrically equal portions*; for this supposition would involve us forthwith in a contradiction similar to that in Item 9, and thus disclose its falsity.

12. We must attribute to a plane extending in all directions to infinity an infinitely great area, and a set of points infinitely many times as great as the set of points in a parallel strip. And
106 just as we credited with equal point sets all parallel strips of the same width, so must we credit all unbounded planes with equal infinite sets of points. For of them too it can be said that they can be determined by figures not merely similar, but also geometrically equal—for example, when we determine each one of them by such three of its points as form a triangle of given size and shape.

13. Every particular unterminated straight line in such an unbounded plane is situated in an exactly similar manner with regard to the parts of the plane on either side of itself, and its situation offers no other conceptually expressible attributes than does the situation of any other straight line of the kind. Yet for all that we are not entitled to say that such a line partitions the plane into two *geometrically equal* portions. For if we could say this of any one straight line *RS*, we should have to say it of any other straight line *R′S′*, which leads to an obvious contradiction so soon as the two lines are taken parallel to one another.

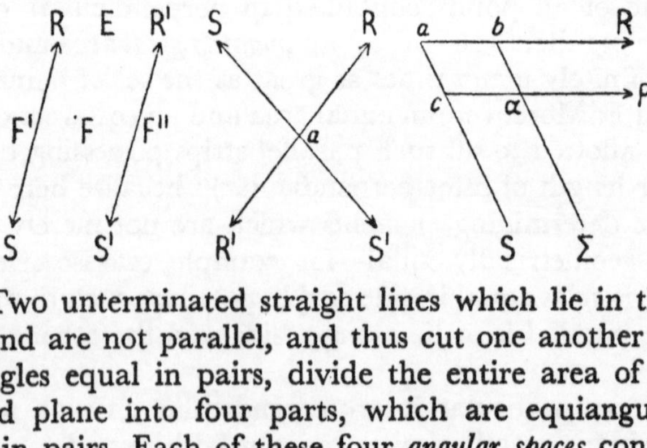

14. Two unterminated straight lines which lie in the same plane and are not parallel, and thus cut one another to form four angles equal in pairs, divide the entire area of the unbounded plane into four parts, which are equiangular and similar in pairs. Each of these four *angular spaces* contains an

infinite set of *parallel strips* like those in Item 11 and stretching in one direction to infinity, all of equal but arbitrary width. After we imagine any particular finite number of these strips to be 107 removed, there remains an angular space subtended by the *same angle* as at first. Yet we are no less unjustified in equating these angular spaces, even when their vertical angles are similar and equal figures, than we formerly were in equating in Items 9 and 11 the legs of these angles, or the parallel strips of which we found their surfaces to be made up. Of the two angular surfaces *RaS* and PαΣ, it is obvious that the former is greater than the latter, although for *b*Σ parallel to *aS* and *c*P parallel to *aR* the angles themselves will be equal.

15. We must judge the solid space included between two unbounded parallel planes (that is, the aggregate of all points contained in perpendiculars dropped from one parallel plane to the other) or what we may call the *unbounded solid slab* to be *infinitely great* irrespective of the thickness (or length of inter-perpendicular). Given two such slabs of equal thickness, however, we must judge their magnitudes, and even the sets of points contained in them, to be equal: and this, by the argument already several times advanced, in Item 8, Item 10 and Item 12.

16. Every particular inter-perpendicular plane *parallel strip* within an unbounded *solid slab* is situated in an exactly similar manner with regard to the parts of the slab on either side of itself, and so is the situation of any other such parallel strip to this or any other solid slab of the kind. Yet for all that, we are 108 not entitled to say that the two portions into which the strip partitions the slab are necessarily *equal in magnitude*.

17. Two unbounded planes which *intersect one another* partition infinite space into four large regions, opposite pairs of which are indisputably *similar* to one another, but are not for that sole reason to be held equal.

18. The like prohibition holds for the solid spaces which two similar (or as we are accustomed to say: equal) solid angles include between their infinitely produced plane sides.

19. Even the two parts into which a single unbounded plane divides the whole of space, similar as they are, may not be considered *geometrically equal* (that is, equal in magnitude) and still less considered as consisting of equal sets of points.

§50

It remains briefly to review the paradoxes to be encountered in the domains of *metaphysics* and *physics*.

In these sciences I advance the following theses: 'The universe contains *no two exactly identical entities, and therefore no two exactly identical atoms* or simple substances;' 'The supposition that there be *composite bodies* in the world necessarily entails the supposition that such simple substances exist;' and 'All these simple substances must be supposed *variable* and constantly in process of variation.' I make these assertions because in my eyes they are truths no less capable of rigorous and clear demonstration than any mathematical theorem. Nevertheless, I am obliged to fear that most physicists will shake their heads on hearing them. For
109 they take pride in setting up only such truths as *experience* teaches them; experience, as they believe, reveals no difference between the minimal particles of bodies, especially of homogeneous bodies; for example, between the minimal particles of gold dug from this or that mine. Granting that experience teaches the compositeness of physical bodies, they go on to urge that absolutely simple and therefore unextended atoms have never yet been perceived by anybody. And granting again how we know from experience that the various kinds of matter (oxygen, hydrogen and so forth) enter now into one combination and now into another and accordingly exhibit now one mode of action and now another, they brand as fictitious the idea of their suffering internal change, and the idea that oxygen for example could gradually change into some other kind of matter.

1. In my view, it is not true that *experience* teaches these assertions. Experience—bare and immediate experience or perception uncombined with purely conceptual truths—tells us no more than the naked fact that we 'have' this or that intuition (Anschauung) or mental representation (Vorstellung). The question whence these representations come, whether they are made under the influence of any object distinct from ourselves, whether indeed they require a cause at all and if so of what kind—is not answered at all by immediate perception (unmittelbare Wahrnehmung) but only by ratiocination from certain purely conceptual truths which reason must adjoin [to experience],

and even this ratiocination follows in most cases a mere rule of
probability: for example, that the redness we are seeing just
now is due to a morbid condition of our eyes, whereas a frag-
rance we are smelling is due to the proximity of a flower. For an
insight, on the other hand, into the fact that some difference or
other must obtain between any two entities, no merely probable
conclusions drawn from experience are needed; on the contrary,
a little reflexion suffices for us to know it with full certainty. If *A* 110
and *B* are to be two entities at all, then precisely on this ground
must the truth hold 'that the entity *A* is not the entity *B*:' a
truth which presupposes the existence of two mental representa-
tions *A* and *B*, of which one represents *A* but not *B*, and the
other *B* but not *A*. And in this very circumstance there reposes
a difference, and an internal difference, between the entities *A*
and *B*. Having seen in this manner that every two entities must
needs exhibit some differences, how can we hold ourselves justi-
fied in doubting of the differences, for no other reason than our
sometimes not perceiving them? After all, no mean sensory
acuity is requisite for doing so, as well as the fulfilment of many
other conditions.

2. There is some truth in the statement that only experience teaches the existence of a number of objects exerting influence upon us, and the *compositeness* of those especially which mediate our intuitions. Yet experience teaches it only after the hypothetical adjunction of certain purely conceptual truths, as for example that different effects can only be brought about by different causes, and the like. No less certain, however, are such conceptual truths as these: that every cause must be something actual; that everything actual must in turn be either a substance, or an aggregate of several substances, or of qualities of one or more substances; furthermore, that qualities cannot exist and be actual apart from the presence of a substance to which they can belong; and again, that aggregates of substances cannot exist unless simple substances exist which can be their members. From this, however, the existence of simple substances follows with strict necessity, and it would be ludicrous to deny the existence of those simple substances just because we do not 'see them.' The absurdity is only enhanced when further reflexion shows that every physical body which is to be perceived by our senses must be

composite, nay more, must be composed of an infinite set of simple parts.

111 3. Another fallacious inference from non-perception to non-existence is incurred when we refuse to admit that all finite substances are subject to incessant *change*. Only too well do we know how variable are the states, the ideas, the dispositions and the powers of our own *souls*, and mere analogy suggests the like for the souls of animals and plants. But the reality of constant change in all substances whatever, even including those which for centuries on end have evinced no change perceptible to us, is something we shall only be justified in accepting on ratiocinative grounds. Those who attack this view, at least as far as what people choose to call *lifeless matter* is concerned, and would question its being true of the *simple parts* of matter, namely atoms, are forced to maintain that all the changes we perceive in this part of creation—for example, when a lump of ice recently solid has now melted, and in an hour's time will evaporate into steam —are nothing else but changes of ubication among the greater or smaller particles of these bodies, unaccompanied by any internal change in the particles themselves. Yet how can it escape anyone that we thus fall into a contradiction? But if no change could take place in the interior of the simple substances themselves, what is there that could possibly bring about changes in their relative ubication? What consequences could ensue upon changes so external as the latter? What purpose could they serve? And by what signs could we come to know them at all? A reasonable answer to all these questions can only be found if the capacity for *change* through mutual influence is attributed to simple substances (more specifically, those which
112 are non-all-perfect) precisely in virtue of their non-all-perfectness, and if their ubications are regarded as those among their determinations which give the reason why, possessing such and such a degree of force, they induce in one another, in such and such a period of time, just this amount of change rather than a greater or a less amount of change. This hypothesis, so evident to the common understanding of mankind, is the only one which dissolves every contradiction in the doctrine of the universe, and in order to reconcile everything it is only necessary to soar above certain *school theses* (Schulmeinungen) which are already almost obsolete.

§51

1. The first *school thesis* we must abandon is that of the earlier physicists who imagined such a thing as *dead* or merely *inert matter*, whose simple parts (when it has any) are indistinguishable and perpetually changeless, with no forces of their own save possibly the so-called *force of inertia*. For whatever is *actual* (wirklich) must really *act* (wirken) and hence possess *powers* (Kräfte) of action. Of course, a substance which is *limited*, and for this reason subject to change, cannot possess the kinds of power which are insusceptible of change in their mode of action, and in particular, cannot possess the power of creation; but it must possess *powers of alteration* exclusively (blosse Veränderungskräfte) and these, moreover, may be either *immanent* like the power of *sentience* (Empfinden) or *transient* like the power of setting in motion (Bewegkraft).

Nevertheless, and for the purpose of judging with sufficient accuracy the result of combining these or those physical bodies, we may be allowed in the future as in the past to begin by representing the case as much simpler than it really is. Instead of the infinite set of forces which in reality are here acting together, we may assume the existence of only a small number. Indeed, we may even imagine non-existent physical bodies, and non-existent 113
qualities inhering in those bodies, with the object of determining what effect they would produce. One prohibition does hold, however: unless the matter has been expressly thought out, we have no right to suppose that results necessary in the hypothetical case will agree up to a point with the results in a real case. As we shall see, many a celebrated paradox must be blamed upon the neglect of this precaution.

§52

2. It is another prejudice of the schools that *science forbids the assumption of any immediate action of one substance upon another*. Now all scientific study would admittedly cease were we content to explain every occurrent phenomenon by dubbing it 'immediately produced:' but this prohibition extends, after all, only to the assumption of an immediate action without previous proof. We should obviously be going too far, and falling into a new and equally disadvantageous error, if we asserted that the action of

one substance upon another could only be *mediate*, and denied immediate action in all cases whatsoever. For how could mediate action come to exist if no immediate action existed? This being sufficiently evident, we have no intention of dwelling on the matter, and only wish to remark how strange it is that so great and so careful a thinker as *Leibniz*, for no better reason than his not knowing of any means whereby simple substances could act upon one another, should have bethought himself of that unhappy *hypothesis of pre-established harmony* which disfigures his otherwise beautiful system of cosmology.

114 §53

3. Intimately bound up with this prejudice, and thus refuted at the same time, is another and much older prejudice that one substance cannot act (that is, cannot act immediately) upon another *at a distance*. I contradict this notion with all possible bluntness, and assert on the very contrary that every action of one (spatially located and thus limited) substance upon another is an '*actio in distans*;' and for the simple reason that at any one instant, each two distinct substances occupy two distinct places and consequently must be *distant* from one another. The apparent contradiction between this and another of my assertions (namely, that space is filled continuously) has already been discussed above.

§54

4. In this, to be sure, we run counter to another prejudice, current in more recent schools, according to which every *chemical* combination involves an *interpenetration* of substances. I deny categorically that any such interpenetration be possible; since, so far as I can see, it is contained in the very *idea of a simple place* (or point) that it be a place capable of affording ubication only to a *single* (simple) substance. Wherever we have two atoms, there do we have two places. It follows too, from our repeatedly given *definition* of space, that the magnitude of the distance between two interacting atoms is the sole determinant of the *magnitude* of the changes they induce in one another during a given space of time. Were it possible for two or more substances to occupy the same place for any space of time how-
115 ever short, then it would be quite impossible to determine the

magnitude of their mutual interaction during that space of time. Even were it but a single solitary instant, still, it would be impossible to determine their condition at that instant.

§55

5. Since the time of *Descartes*, the schools have seen the rise of yet another prejudice. From the best of motives, no doubt, Descartes deemed it impossible to overestimate the difference between *thinking* and *non-thinking* substances (*spirit* and *matter*, as he called them) and thus came to make an assertion which comes as a shock to the minds of people at large, and is indeed almost unthinkable. His assertion was, not merely that spiritual beings cannot be regarded as extended and consisting of parts, but over and beyond this: that spiritual beings cannot be regarded as found 'in space' in any way whatever, and hence not regarded as filling even a single solitary *point* of space with their presence. Later on, *Kant* went so far as to declare that space and time are neither of them anything more than mere forms of our sense-activity for which there is no corresponding thing-in-itself. In view of this, and of his setting up in opposition to one another an *intelligible world* and a *world of sense and spirit*, it is hardly to be wondered at that, in Germany at least, the prejudice in favour of the unspatiality of spiritual beings should have taken root so deeply and held its ground in our schools down to the present day. The grounds on which I believe myself to have attacked this prejudice must be sought in other writings of mine, notably in my '*Wissenschaftslehre*' and in my '*Athanasia.*' My opinion that all created substances must, for a reason common to them all, be immersed in space as well as in time, and that all differences in their powers are differences of degree alone, has at least one recommendation which all are bound to concede: it is at any rate the simplest of all the opinions hitherto broached and known.

§56 116

6. If this view is taken, the great paradox hitherto seen in the *linking together of spiritual and material substances* is automatically eliminated. The question of how matter can influence spirit, and spirit influence matter, the two being so utterly disparate, has been proclaimed as a mystery inaccessible to human investigation. The above view leads, however, to the conclusion that

this mutual influence must be *immediate*, at least in part, and to this extent therefore neither secret nor occult. Yet it shall not be hereby denied that many aspects of this mutual influence are well worth our knowledge and further research, especially those in which the mediation is *effected by organisms*.

§57

7. Whereas some of our ancestors imagined *substances without forces*, attempts have been made in more recent times to construct the universe conversely out of *forces without substances*. The false formulation of the idea of substance, that a substance be a *mere* aggregate of forces, was doubtlessly occasioned by the fact that no substance can manifest itself to us otherwise than through its effects, and thus through its *power* to produce effects. Furthermore, the grossly material metaphor inherent in the etymology of the words 'substance,' 'substrate,' 'subject,' 'carrier' (Träger) appeared to offer a clear proof that the generally received doctrine, according to which the existence of a substance necessitates after all a special *Something* in which its forces can inhere as *qualities*, is nothing but an illusion of the senses; for here there is certainly no need of a *carrier* (Träger) or
117 a *substratum* (Unterlage) in the strict sense of the term. Yet are we really obliged to rest content with this sensory interpretation? Every Something whatever, even the bare *idea* of *nothing*, must surely be considered as an object claiming not just one, but a whole aggregate of infinitely many qualities. Do we then for such a reason as this look upon every Something whatever as a carrier in the strict sense of the term? Certainly not! But if we think of Something, and delimit that Something as having to be something *actual* and no mere quality of some other actual Something, we shall then be subsuming it under the idea of a *substance* according to the correct definition of the term. And besides the single uncreated Substance there exists an infinite set of created ones. Following current linguistic usage, we give the name 'forces' to all those qualities in these substances which must be presupposed as affording the proximate (or immediate) cause of Something Else: whether that Something Else lies within or without the substance which is effecting it. If we had a force which was not a quality inherent in some substance, then —since in its function of cause it would have to be actual, and

thus be something actual yet *not* inherent in something else also actual—it would precisely for this reason be more than a mere force: we should have to call it rather a *substance* in its own right.

§58

The fact that in all God's creation no *degree of being* is the *highest* and none the *lowest*; the further fact that in every degree however high, and at every instant however early, creatures have existed which have risen to that degree by their rapid progress; and the converse fact that in every degree however low, and at every instant however late, creatures will exist which despite continual progress have only then attained that degree: these paradoxes need no further justification beyond what has already been said in §38 and its sequel on a similar state of affairs with regard to space and time.

§59 118

A certain other *paradox*, however, will sound much more scandalous. It is my thesis that: 'Although the *whole infinite space of the universe* is everywhere and at all times so completely filled with substances that no single point is either for a single instant unoccupied by a substance, or occupied by two or more substances; nevertheless, the various parts of space are filled at various times to an infinite number of different *degrees of density*; so much so that the very same set of substances, which at this particular moment fills this particular cubic foot, could be distributed another time throughout a space one millionfold larger, without any point in this larger space standing empty; and distributed yet another time throughout a space one millionfold smaller, without any point in this smaller space needing to accommodate two or more atoms.'

I am fully aware that what I here assert has hitherto been looked upon by most physicists as absurd. Their belief in the irreconcilability of a continuous filling of space with the fact that *bodies have different densities* is the very reason why they assume a sort of *porosity* to be a general property of matter even in cases, such as those of gases and the ether, where no observational evidence exists at all. In these pores, the greater among which are supposed to be filled with gases—and thus, strictly speaking, only in the never-observed pores of fluids—physicists

have assumed down to this day the existence of their so-called '*vacuum dispertitum*,' namely: certain empty spaces of such a number and extent that hardly the billionth part of a region filled with ether contains genuine matter. Despite this, I hope that all who have taken into proper consideration what was said in and
119 after §20 will see clearly enough for what reasons there is no sort of impossibility in one and the same (infinite) set of atoms being distributed, now in a larger region without a single point standing solitary there, now in a second and contracted region without a single point requiring to absorb two atoms.

§60

But little opposition is likely to be encountered, at this point in our argument, by the thesis—already maintained by the older metaphysicians in their theory '*de nexu cosmico*'—that every substance in the universe is engaged in mutual interaction with every other substance in the universe, subject to the limitation that the changes they work in one another become the smaller, the farther apart they stand, and to the limitation that the *total result of the influence of all substances on any one substance* is (divine intervention apart) a change subject to the *law of continuity*; seeing that a departure from that law would call for a force *infinitely great* in comparison with any which continuity would call for.

§61

Easily as the theory of *dominant substances* (*herrschende Substanzen*) was deduced from pure concepts even in the first (1829) edition of the '*Athanasia*,' people will find paradoxes here as well, wherefore it becomes needful to devote a few words to their consideration.

In the passages in question, I start from the notion that, some finitely great difference being admitted as necessary between
120 two substances at each particular instant, there must exist at each particular instant some substances whose forces have grown large enough to exercise a predominance over all other substances, at least in some sufficiently small vicinity of themselves. It would be a mistake, and a mistake which threw on my hypothesis the suspicion of harbouring a contradiction with itself, if anyone imagined that a dominant substance of this kind needed to possess forces infinitely greater than those of the

dominated substances. This is by no means the case. For suppose we have a region of finite extent, say a sphere, and within it, say at the centre, a substance whose forces exceeded those of the other substances [in the sphere] in a finite ratio, as would happen, for example, if each of the latter were, say, only half as strong as the former. This being supposed, it is quite true, to be sure, that the total effect of these infinitely many weaker substances could outweigh that of the single stronger one infinitely many times, namely in cases where their action converges; which happens, as we shall soon see, with their tendency to draw near to a central body. On the other hand, cases can occur and must occur in which those [various weaker] forces do not happen to converge. In particular, if we attend solely to the influence exercised by each substance in our region upon the others, and received by each from the others, we must say as a rule that this mutual influence will be stronger in the direction from the stronger substance, and stronger in proportion to its greater strength. In our chosen example, the substance assumed to be at least [*sic*] twice as strong as each of its neighbours will act at least twice as strongly upon them as they re-act upon it. This it is, and this alone, that we are thinking of when we say that the one substance *dominates* all the others.

§62 121

If that be the situation, someone will perhaps remark, then we must find a dominant substance not only in certain regions of space, but in every region of space however small, and indeed in every possible aggregate of atoms: for in every aggregate of several atoms, there must surely be a strongest, and as surely a weakest? Now I hope that none of my readers need be taught that this can be inferred only for finite sets, and that where we have an infinite set, it is possible for every member to have another member higher than itself, and another member lower than itself, in spite of the circumstance that no single member rises above a given finite quantity, and no single member falls below another given finite quantity.

§63

By their very definition these *dominant* substances can only occur to a finite number in any one finite region of space, and they are surrounded sometimes by a larger, sometimes by a smaller envelope of merely auxiliary (dienende) substances. United in clusters of finite size, it is they which constitute what we call the various *bodies* in the universe, whether gaseous or fluid or solid or organic or any other kind. To all the remaining world-stuff, which fills the rest of space without containing privileged atoms (augezeichnete Atome) I assign by contrast the name of *ether*. This would not be an appropriate place to set forth the advantages of my present hypothesis in furnishing a very easy explanation of many a phenomenon hitherto explained imperfectly or not at all. The purpose of this book only allows me to offer a few hints for the clearing up (21) of apparent contradictions.

122 If all created substances differ from one another only in the *degree* of their forces, if we must consequently attribute to each of them a certain degree of '*sensitivity*' (Empfindung) be it ever so small, and if each pair of substances act upon one another, then it is the easiest thing in the world to understand how, for a particular pair of any kind, and most of all for a pair of dominant substances, *not every arbitrary distance will be equally acceptable to them*, or equally beneficent for them: because the magnitude of the influence they exercise, or of the influence to which they submit, depends upon the magnitude of their distance. If the distance which they happen to have is greater than is acceptable to one of them, then a tendency to shorten that distance will come into play—which we call an *attraction*; and in the opposite case, a *repulsion*. Neither of the two should be thought of as necessarily reciprocal, still less as necessarily accompanied by an actual change of position; but what we are entitled to suppose is, that given two substances in the universe, there exists one distance so great that for it, and all others greater still, a reciprocal attraction takes place; and another distance so small that for it, and all others smaller still, a reciprocal repulsion. Greatly as these two limiting distances for attraction and repulsion may vary in the course of time, both with the nature of the two given substances themselves and with the nature of all the substances in their vicinity, this much is certain: that, all other circumstances remaining similar, the influence exerted by two substances on one another must decrease as their distance increases—if only for the reason that

the number of those which could find room at the same distance, and thus lay claim to the same influence, grows with the square of the distance.

Moreover, since the preponderance of a dominant over an auxiliary substance never rises above a finite magnitude, where- 123
as the number of the latter in a given space exceeds the number of the former infinitely many times, it is easy to see that the amount of attraction which all the substances included in a given space exert on an atom lying outside that space will, once the distance becomes sufficiently large, be approximately the same as if that space contained no privileged substances at all, but only an equal number of common atoms.

In combination with what went before, this leads to the important conclusion that '*as soon as two bodies are distant by a sufficient amount, there is exerted between them a force of attraction proportional to the sum* [sic] *of their masses* (the number of their atoms) *and the inverse square of their distance.*' In our days, no physicist and no astronomer denies that this law is followed in the universe; but people appear very seldom to have noticed how hard it is to reconcile this with the common view of the nature of the elementary particles of the different kinds of body. If the accounts hitherto given were really true; if the 55 or more *elements* (einfache Stoffe) which chemists have learnt to know on this earth made up the mass of all existing bodies in such a way that each and every one of them were nothing but a mere aggregate of atoms, sometimes all of this kind, sometimes all of that kind, sometimes of several kinds mixed; if gold were a mere aggregate of gold atoms alone, and sulphur a mere aggregate of sulphur atoms alone: then let anyone try to explain to me how it comes to pass that kinds of matter which differ so greatly in their forces, and particularly in the degree of their mutual attraction, can for all that be absolutely equal in *weight*, that is, how it happens that their weights are proportional to their masses. The fact that they are so is proved at once by the well- 124
known experiment in which spheres made of any chemical substance we please behave, if only their weights are equal, exactly as bodies of equal mass must behave—therefore, for example, bringing one another to rest when the velocities are equal and the influence of elasticity is either eliminated or taken account of.

If we assume instead that all bodies consist of nothing but an infinite amount of ether together with a number of privileged atoms which in comparison therewith is vanishingly small, and assume that the forces of these privileged atoms exceed those of an ether atom only a finite number of times, we understand how it is that the force of attraction experienced by these bodies as coming from the entire terrestrial sphere cannot be noticeably increased by so small a number of privileged atoms, and so understand how it is that their weight must be proportional to their total mass.

Yet even now physicists are not wanting who regard *caloric matter* (Wärmestoff), which is fundamentally identical with what I myself call the ether, as a fluid present in all bodies and impossible completely to drive out of them. Had they then not been so unfortunate as to get hold of the notion that this caloric matter is *imponderable*, and had they lifted their minds high enough to see that over and beyond its caloric matter each individual body contains only a vanishingly small number of other atoms: then it would very soon have become clear to them that just this caloric matter of theirs is what determines the weight of every individual body. And how near they were to all this, when they sometimes postulated that the non-caloric atoms be infinitely distant from one another in comparison with their own diameters!

125 §64

The dominance of a privileged substance over its immediate vicinity is easily understood to consist, if in nothing else, then at least in a certain stronger attraction on its neighbouring atoms, in consequence of which these neighbouring atoms lie closer to it and to one another than they would otherwise do. For this very reason, they have a tendency to withdraw a little from the point of attraction and from one another as soon as an opportunity occurs, and thus a tendency to *repel* one another. Many observations point to this, but in order to explain it, people have needlessly assumed the existence of a *primitive* force of repulsion between the particles of the *ether*.

§65

The same circumstance affords an easy proof of the proposition which I laid down in my '*Athanasia*,' to the effect that: '*No*

privileged substance ever suffers so great a change in its envelope as not to retain some fragment however small of its original immediate vicinity.' No one, to be sure, could fear lest a privileged substance *a* be robbed of its immediately surrounding ether atoms while all the nearest other privileged substances *b*, *c*, *d*, *e*, . . . are keeping fixed distances; but we could be afraid of it when some or all of them moved away. Yet even then, only a fraction of the ether particles surrounding *a* could be captured by the fleeing substances *b*, *c*, *d*, *e*, . . . and another fraction, from among those nearest to *a*, must always remain behind. At the same time, we not only admit, but even declare it to be necessary, that the latter fraction would have to expand into a roomier space. In certain circumstances indeed, ether atoms 126
would have to flow in from sundry remote regions and penetrate into those spaces which the all-too-wide dispersal of the substances *b*, *c*, *d*, *e*, . . . has left filled with a comparatively spongy (locker) ether. There is no reason why the ether coming from a distance should push away and replace the ether immediately surrounding *a*. Instead of completely driving away the ether surrounding *a*, the affluent ether must hinder its further spread and compress it until its density achieves equilibrium with the forces of attraction exerted by all circumambient atoms.

§66

Closely following upon this, many questions now find an answer which would have been deemed paradoxical, had the preceding exposition not provided the requisite information. To this category belongs the problem concerning the *limiting surfaces of physical bodies*: where exactly does one such body end and another begin?—Now I define the *limiting surface* (Gränze) of a body as the aggregate of all the *extreme* (äusserst) ether-atoms which still *belong to it*. And by 'belonging to it,' I mean that they are more strongly attracted by the privileged atoms of the body in question than they are by any other dominant atoms present in the vicinity, in such wise that if the body should alter its position relative to that vicinity, for example by moving away, these atoms would still follow in its wake; perhaps not with the same speed, to be sure, yet without any separation taking place, and without the intrusion of foreign atoms. Once this definition of a limiting surface is laid down, it soon becomes evident that the limiting surface of a body is something very mutable, indeed, something that is almost always in process of change, every time that any alteration occurs partly in it and partly in the neighbouring bodies; because, as may readily be conceived, every alteration of that sort could bring about manifold changes both in the amount and in the direction of the attraction which is experienced, not only by the auxiliary, but even by the dominant atoms of the body. Sundry particles of this quill, for instance, were a moment ago more strongly attracted by its remaining mass than by the surrounding air, and thus belonged to it; but now they are more strongly attracted by my fingers than by the mass of the quill, and are thus wrenched away from it. A closer consideration shows further that many bodies are at certain places altogether devoid of limiting atoms; none of their atoms can be described as the *extreme* ones among those which still belong to it and would accompany it if it started to move. Whenever, in fact, we have two neighbouring (22) bodies, and a place where one of them possesses an extreme atom ready to accompany it, the other of the two is for this very reason unable to possess an extreme atom; since all atoms situated behind the extreme atom of the one body already belong to the other body.

§67

Yet another question receives its answer from this source: the question whether and when two bodies stand in immediate *contact*, or whether they are separated by an intervening space. If I may do what strikes me as most expedient, and define the contact of two bodies as taking place when the extreme atoms of the one, as understood in §66, together with certain atoms of the other, form a continuous extension—then it will not admit of denial that in quite a large number of cases bodies really are in mutual contact. This is not only true when at least one of them, and still more when each of them, is in the fluid state, but even when both of them are solid: provided only that we get rid, by intense squeezing or in some other way, of the air usually adherent to them in normal terrestrial conditions. If on the other hand two bodies are not in contact, then some other body,
128 or at least pure ether, must lie between them; because absolutely empty space is impossible. We are consequently in a position to

assert that, strictly speaking, every body is, on every side of it, either in contact with some other body, or in the absence of any other body, in contact with pure ether.

§68

Turning to the subject of the various types of movement in the universe, our view that no portion of space is empty might seem to exclude all kinds of movement save those in which any whole mass in simultaneous movement forms a single re-entrant extended figure, and each part of the mass enters only into such places as have just been occupied by another part of the same mass. Those who remember what was said in §59 about the different degrees of density with which space can be filled will, however, understand that many other movements are possible and necessary. For a reason so obvious that I have no need to mention it at all, there is one particular kind of movement which must take place almost incessantly, not only with ether-atoms, but with nearly all privileged atoms as well: to wit, *oscillatory* movement. In the next place after this, and especially with *solid* bodies, *rotatory* movement must be very common. In our opinion the axis of rotation must always be a material line; but the problem of how the rotatory movement must be conceived, and how half a revolution brings atoms from one side of the axis to the other without their cutting adrift, will cause confusion only in the minds of those who forget that in the interior of a continuum, no less than outside a continuum, every atom stands at a certain distance from every other atom, and is therefore able
to revolve round the other atom without either itself breaking 129
loose, or having to drag the other atom round with it. In the case of a simple spatial entity, the latter contingency, a rotation about itself, would be self-contradictory.

§69

The description of a perfectly straight or perfectly circular path is something which we do not venture to assert of any single atom in the universe, dominant or common, at any single instant; in fact, given the infinite set of perturbations which each atom undergoes under the influence of all the others, the improbability of such a path becomes infinitely great. At the same time, we have no right to declare such a path to be impossible in itself. What we may assert is, for example, that the path can only be a *broken line* when the velocity of the atom gradually decreases towards the end of the segment *ab* and becomes zero at the point *b* itself, afterwards increasing from zero upwards at every instant subsequent to the arrival of the atom at *b*, excepting only the case where a finite period of rest at *b* is to be interpolated.

With certain other lines the situation is different, and particularly with the logarithmic spiral. Even when we abstract from all perturbations of external origin, it is self-contradictory for an atom to start at a given point on that spiral and traverse the centripetal branch in a finite space of time, and still more absurd to demand that the moving atom should ultimately arrive at the pole. Let us take and prove only the case where it moves with uniform arcual velocity, first supposing everything else at rest. It does not take us long to see that the spiral move-
130 ment can be resolved into two components, one of them along the radius vector with constant velocity towards the pole, and the other an angular velocity round that pole which increases constantly and must surpass any arbitrary amount in order that the atom may come arbitrarily near to the pole. It is quite certain, therefore, that there is no force in nature able to impart this velocity to it, and all the more certain that no force can impart it to a whole three-dimensional mass of atoms so as to make every one of the atoms traverse the infinitely many whorls right into the pole, and this in a finite space of time. But even if all this could be obtained, would it really be possible to say that the pole had been reached? I for my part do not think so. The one-dimensional extension consisting of the pole together with all the points indubitably on the spiral does indeed form a continuum—because then every point does have a neighbour at any chosen distance however small—yet another requirement for the path of a moving atom remains unfulfilled, the requirement namely that the path should have at every one of its points one or several definite directions. This is well known not to be true at the pole [of a logarithmic spiral].

In this context, the teasing question finally occurs, whether our opinion about the infinitude of the universe leaves room for a translatory motion of the entire cosmos in any definite direction, or for a rotatory motion round some fixed axis or centre. The reason for our negative answer is not that places would be wanting into which the atoms could move, for this by itself would not compel us to declare the two suggested movements impossible. The true reason for their impossibility is, that causes (forces) are wanting which could bring such a movement about. For as a source of the occurrence of such a movement, we are unable to think, either of a *physical ground*—an arrangement 131 which, being a simple consequence of purely theoretical and conceptual truths, is unconditionally necessary, or of a *moral ground*—an arrangement which, being present only because God brings about every event advantageous to the well-being of His creatures, is only conditionally necessary.

§70

Let us bring these considerations to a close with two paradoxes whose celebrity is chiefly due to *Euler*.

We begin with the question how an atom *a* moves when attracted in inverse proportion to the square of the distance by a force situated at *c*. *Boscovich* had already drawn attention to the fact that one answer is obtained when we project the atom from *a* and seek the limit of the resulting elliptic motion when the velocity of projection tends to zero, but a different answer when we start with no such fiction but consider the problem directly. If a velocity perpendicular to *ac* had been imparted to the atom, by projection or otherwise, and if we abstract from all resistance in the medium, the atom would have had to describe an ellipse with one focus at *c*. Now if the lateral velocity of projection diminishes infinitely, the minor axis of the ellipse does the same, whence *Euler* concluded that for a lateral velocity zero the result must be an oscillation to and fro between the points *a* and *c*, this alone being the movement into which the elliptic movement can pass over without violating the law of continuity. Other people, and *Busse* in particular, considered it absurd that an atom moving with a velocity which increased to infinity as it 132 approached nearer to *c* should be checked in its course and driven back in the opposite direction; for no such hypothesis as that of an impenetrable atom fixed at *c* and preventing transit through *c* had been laid down at all. These other people therefore maintained that the atom must rather continue to move in the direction *ac* up to *c* and beyond, though now with decreasing velocity, eventually reaching the end of the segment $cb=ca$ and then returning from *b* to *a* and repeating the process indefinitely.

My own view is that in this matter *Euler*'s appeal to the law of continuity can decide nothing. The kind of continuity which can be proved to govern the changes in the universe (in the growth or waning of the forces of individual substances) is violated by the phenomenon in dispute, neither when the oscillation takes place between the bounds *a* and *b*, nor when it takes place between *a* and *c*. The real offence against the law of continuity was already committed, beyond hope of justification, when a force of attraction was supposed to increase infinitely. On this ground alone, we have no right to be astonished at contradictory conclusions following from contradictory premises.

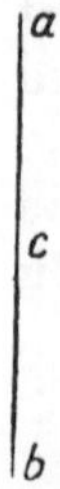

This shows, moreover, that *Busse*'s solution of the problem is just as wrong as *Euler*'s, because it assumes something intrinsically impossible, to wit: an infinite velocity at *c*. If this error is corrected, if we assume instead that the velocity of the atom follows a law which keeps it always finite, and if we remember
133 that the movement of an individual atom can never be discussed apart from the medium in which it moves and from the greater or smaller number of atoms moving with it: then quite a different solution is obtained, whose details do not need to be set forth at this point.

The *second* of the (Eulerian) paradoxes about which we would like to add a few remarks is one concerning the *motion of a pendulum*. Everyone knows that the quarter-period of a simple pendulum of length *r* over an infinitely small amplitude is calculated at $\frac{\pi}{2}\sqrt{\frac{r}{g}}$, whereas the time of fall down the corresponding chord, usually considered equally long, turns out to be $\sqrt{2}\,.\sqrt{\frac{r}{g}}$. The reason why *Euler* thought this paradoxical was simply his false notion of the *infinitely small*, which he equated to zero. In reality, however, infinitely small arcs are just as non-existent as infinitely small chords, and the statements which mathematicians make about their so-called 'infinitely small arcs and chords' are statements which they only prove for arcs and

chords that we can take as small as we please. The right meaning of the above two equations can only be as follows: (i) the quarter-period of a pendulum can be brought as near as we please to the quantity $\frac{\pi}{2}\sqrt{\frac{r}{g}}$ by taking the arc of oscillation sufficiently small; and (ii) the time of falling down a chord approaches by the same means as near as we like to $\sqrt{2}\,.\sqrt{\frac{r}{g}}$. The fact that these two quantities are unequal, and that arc and chord differ with respect to the time of fall no matter how small they are taken, is no more startling than many another dif- 134
ference between them. As long as both are in existence, no one expects these differences to disappear—such differences, for example, as that the arc always retains a curvature whose magnitude we can assess at $1/r$, whereas the chord is always straight and has no curvature at all.

REFERENCE NOTES TO TRANSLATION

(1) Besides being translated literally as they occur, Bolzano's references to mathematical and philosophical literature are collected and supplemented on pages 182–184 of the present volume.

(2) Bolzano writes 'erst durch,' not merely 'durch.'

(3) *Teile* often meaning *individual members*.

(4) Printed thus in the original, and not in the more minutely analytical form $A+(B+C)=(A+B)+C$.

(5) Bolzano, or Přihonský for him, writes ambiguously: 'Verhältnis.'

(6) In the text, 'Grösse (Grossheit).'

(7) The translation uses subscript capitals N in place of the original capitals N with *vertically* superscript 0 and 1 and 2, respectively n.

(8) In the text, the left side was: $S_1 - S_2$.

(9) The 'absence,' not the 'representation of the absence,' in the text: 'als blosse Abwesenheit.'

(10) The notation of the original was '0 $(A) = 0$.' which the errata corrected to '$0(A) = 0$.'

(11) Printed in the original as follows:

$$\frac{M}{N} > \frac{p}{q} < \frac{p+1}{p},$$

the last denominator being corrected on its subsequent occurrences, but not the mixture of linguistic and notational syntax. In the German text, either the author or the posthumous editor here lapses into an anacoluthon, the original being: 'Stehet die eine derselben, z.B. M, in einem irrationalen Verhältnisse zum Maasse N, so kann es sich allerdings treffen, dass wir bei der gewöhnlichsten Weise des Messens, welche zu jeder beliebigen auch noch so grossen Zahl q eine andere p von der Beschaffenheit sucht, dass . . . sei; und es kann sich fügen, dass auch . . .'

(12) The original edition did not use brackets in the modern style, but printed simply Fx and ϕx.

(13) In the text: 'für jede hinlänglich kleine Entfernung.' That

'at' is meant and not 'within,' is confirmed by §38 and Section 3 of §41.

(14) The terms 'aggregate' and 'set' and 'multiplicity' were distinguished in §4.

(15) The original title is: *Grundriss der gesammten höheren Mathematik.*

(16) The errata ask this to be read in place of

$$\pm n\pi \pm \frac{\pi}{2}.$$

(17) This spelling of a diversely transcribed name is the one used by Bolzano.

(18) The original title is: 'Anfangsgründe der höheren Analysis.'

(19) By a 'natural spiral' was then meant a logarithmic one whose constant angle of intersection with the radius vector is $\pi/4$. See, for example, §43, page 427, volume IV of the mathematical dictionary cited by Bolzano in §12 above.

(20) The original printed: $bR=(ba+aR=bS=(aS-ab)$ and corrected this in the Errata to $bR(=ba+aR)=bS(=aS-ab)$. The translation adopts a verbal paraphrase so as the better to bring out the original argument without disturbing the original non-orientation.

(21) The sense-destroying misprint 'aufgestellt' (set up) was corrected in the Errata to 'aufgehellt.'

(22) 'Nachbarlich' in the text, though 'tangent' seems to be meant; consult the opening of §67. In Bolzano, the word 'Atom' is masculine in gender.

Imprint of the original (1851) *edition*:

Printed by C. E. Elbert in Leipzig.

Imprint of the fascimile (1889) *edition*:

Anastatic reprint by A. Dannenberg, Berlin, N.

The fascimile substitutes its own date on the title-page and does not reproduce the original imprint.

SELECTED BIBLIOGRAPHY

BIBLIOGRAPHICAL WORKS

(B1) BERGMANN, H.: *Das philosophische Werk Bernard Bolzanos, mit Benutzung ungedruckter Quellen kritisch untersucht, nebst einem Anhange: Bolzanos Beiträge zur philosophischen Grundlegung der Mathematik*. Halle, 1909. There is an eight-page bibliography to date on pp. 212–219.

(B2) ENYVVÁRI, E.: *Bolzano-Bibliographie*. Budapest, 1912. Not seen by the Translator.

(B3) E. FRANZIS announced (B60, page xx) that he and E. WINTER were preparing a complete catalogue of all Bolzano's works and of the literature on Bolzano.

BIOGRAPHICAL WORKS

(B4) FELS, H.: *Bernhard Bolzano, sein Leben und sein Werk*. Leipzig, 1929, x+109.

(B5) FELS, H.: 'Die Verurteilung Bolzanos, dargestellt auf Grund der Prozessakten.' *Philosophisches Jahrbuch der Görres-Gesellschaft*, 44 (1931), 419–447.

(B6) FESL, M. J.: *Lebensbeschreibung des Dr. B. Bolzano mit einigen seiner ungedruckten Aufsätze und dem Bildnisse des Verfassers*. Sulzbach, 1836, lvi+272. Pages 1–88 bring the Autobiography, dated 28 May 1831. Also: Vienna, 1875.

(B7) GOTTHARDT, J.: *Das Wahrheitsproblem in dem philosophischen Lebenswerke Bernhard Bolzanos*. Dissertation Münster, 1918; Trier, 1918, xli+133+33. One chapter in full, with a detailed analysis of the rest, covering much biographical matter and interpretation.

(B8) WINTER, E.: *Bernhard Bolzano und sein Kreis, dargestellt mit erstmaliger Heranziehung der Nachlässe Bolzanos und seiner Freunde*. Leipzig (Preface 1932), o+288.

BY BOLZANO ON LOGIC

(B9) *Dr. Bolzanos Wissenschaftslehre. Versuch einer ausführlichen und grösstenteils neuen Darstellung der Logik, mit steter Rücksicht auf deren bisherige Bearbeiter*. Sulzbach, 1837. Volume 1, xvi+571; volume 2, viii+568; volume 3, viii+575; volume 4, xx+683.

Zweite Auflage, herausgegeben von W. Schultz, Leipzig, 1929–1931. Volume 4 of this edition adds an index of authors, works and passages.

(B10) *Bolzanos Wissenschaftslehre und Religionswissenschaft in einer beurteilenden Uebersicht. Eine Schrift für alle, die dessen wichtigste Ansichten kennen zu lernen wünschen*. Sulzbach, 1841, 0+231. Pages 19–118 bring a summary of the *Wissenschaftslehre*.

(B11) 'Ueber einige logische Fragen, angeregt durch Exner's "Nominalismus und Realismus."' Read in 1842, printed in KBGW, (5) 2 (1843), 71–78. For Exner's own paper, see there pages 409–424.

(B12) *Etwas aus der Logik*. Manuscript in the National Museum at Prague, described by E. Winter (B64, 17–18).

BY BOLZANO ON MATHEMATICS

(B13) *Betrachtungen über einige Gegenstände der Elementargeometrie*. Prague, 1804, 63 pp.

(B14) *Beyträge zu einer begründeteren Darstellung der Mathematik*. Prague, 1810, 152 pp. Only one instalment appeared.

(B15) *Der binomische Lehrsatz und als Folgerung aus ihm der polynomische und die Reihen, die zur Berechnung der Logarithmen und Exponentialgrössen dienen, genauer als bisher erwiesen*. Prague, 1816, 144 pp.

(B16) *Die drey Probleme der Rectification, der Complanation und der Cubirung, ohne Betrachtung des unendlich Kleinen, ohne die Annahme des Archimedes und ohne irgend eine nicht streng erweisliche Voraussetzung gelöst; zugleich als Probe einer gänzlichen Umgestaltung der Raumwissenschaft allen Mathematikern zur Prüfung vorgelegt*. Leipzig, 1817, 80 pp. Reviewed in some detail in the *Allgemeine Literaturzeitung*, Halle und Leipzig, 1819, Band III, 180–184.

(B17) *Rein analytischer Beweis des Lehrsatzes, dass zwischen je zwei Werten, die ein entgegengesetztes Resultat gewähren, wenigstens eine reelle Wurzel der Gleichung liege*. Prague, 1817, and KBGW, (3) 5 (1818), in ninth pagination, 60 pp.

Berlin, 1894. Wissenschaftliche Klassiker in Faksimiledrucken, Band VIII.

Leipzig, 1905. Edited by P. E. B. Jourdain in *Ostwalds Klassiker der exakten Wissenschaften*, Heft 153.

(B18) *Ueber den eigentümlichen Ideengang, der meiner Ansicht nach bei einer streng wissenschaftlichen Darstellung der Mathematik befolgt werden müsste.* Read in 1842, not in KBGW.

(B19) *Von Haltung, Krümmung und Schnörkelung bei Linien sowohl als Flächen, samt einigen verwandten Begriffen.* Read in 1843–44, not in KBGW.

(B20) *Dr. Bernard Bolzanos Paradoxien des Unendlichen, herausgegeben aus dem schriftlichen Nachlasse des Verfassers von Dr. Fr. Přihonský.* Leipzig, 1851, xii+134.

Berlin, 1889. Wissenschaftliche Klassiker in Faksimiledrucken, Band II.

Leipzig, 1921. Edited by A. Höfler, annotated by H. Hahn in the Meinersche Philosophische Bibliothek, Band 99, ix+156.

(B21) *Allgemeine Theorie der Funktionen.* Read in 1846, not in KBGW.

(B22) *Funktionenlehre.* Herausgegeben und mit Anmerkungen versehen von K. Rychlík. Prague, 1930, xx+183+24+vi. Spisy Bernarda Bolzana, Bernhard Bolzanos Schriften, herausgegeben von der Böhmischen Gesellschaft der Wissenschaften, Band I.

(B23) *Zahlentheorie.* Herausgegeben und mit Anmerkungen versehen von K. Rychlík. Prague, 1931, 57+7. Spisy Bernarda Bolzana, Bernhard Bolzanos Schriften, herausgegeben von der Böhmischen Gesellschaft der Wissenschaften, Band II.

(B24) *Ueber die Rechnung mit unendlichen Zahlenausdrücken.* Read in 1846, not in KBGW.

BY BOLZANO ON PHYSICS

(B25) *Aphorismen über die Physik.* Read in 1840–41, not in KBGW, and withdrawn by Bolzano.

(B26) *Versuch einer objektiven Begründung der Lehre von der Zusammensetzung der Kräfte.* Prague, 1842, and KBGW, (5) 2 (1843), 425–464.

(B27) *Versuch einer objektiven Begründung der Lehre von den drei Dimensionen des Raumes.* Prague, 1843, and KBGW, (5) 3 (1845), 201–215. Said in the exordium to be based on attempts beginning in 1815 and since constantly revised.

(B28) *Versuch einer sehr einfachen Erklärung des Phänomens der Aberration des Lichtes.* Read in 1845, not in KBGW, edited posthumously by E. Wewerka, Prague, 1853.

(B29) 'Ein paar Bemerkungen über die neue Theorie in Herrn Dopplers Schrift über das farbige Licht der Doppelsterne und einiger anderen Gestirne des Himmels.' *Poggendorffs Annalen*, 60 (1843), 83–88.

(B30) 'Christian Dopplers neueste Leistungen auf dem Gebiete der physikalischen Apparatenlehre, Akustik, Optik und optischen Astronomie.' *Poggendorffs Annalen*, 72 (1847), 530–555.

BY BOLZANO ON OTHER SUBJECTS

Numerous writings of Bolzano on subjects of religion, philosophy, theology, aesthetics, sociology, political philosophy and so forth are listed by H. Bergmann (B1, 122 and 212–219). Only the following works call for mention in this place.

(B31) *Athanasia, oder Gründe für die Unsterblichkeit der Seele. Ein Buch für jeden Gebildeten, der hierüber zur Beruhigung gelangen will.* Zweite Ausgabe, Sulzbach, 1838, pp. i–xii and 16–460. The first edition appeared in 1827, but the second is usually quoted. This work contains, for example, Bolzano's philosophy of substance.

(B32) *Ueber den Begriff des Schönen.* Prague, 1843, and KBGW, (5) 3 (1845), 1–92.

(B33) *Ueber die Einteilung der schönen Künste.* Prague, 1843, and KBGW, (5) 6 (1851), 133–178. This title and the previous one are based on several papers read between 1842 and 1847, with varying and overlapping subjects.

(B34) *Vom besten Staate. Nach den Manuskripten des Nationalmuseums in Prag.* Herausgegeben von A. Kowalewski, Prague, 1932, xxviii+134. Spisy Bernarda Bolzana, Bernhard Bolzanos Schriften, herausgegeben von der Böhmischen Gesellschaft der Wissenschaften, Band III.

(B35) *Paradoxien in der Politik.* Aus Bolzanos Nachlass herausgegeben von. W. Stähler. Münster in Westfalen, 1933, lxv+173. Contains 'Vom besten Staate,' on pages 1–137, and another work on the improvement of pauper institutes.

ON BOLZANO THE LOGICIAN

(B36) DUBISLAV, W.: 'Bolzano als Vorläufer der mathematischen Logik.' *Philosophisches Jahrbuch der Görresgesellschaft*, 44 (1931), 448–456.

(B37) FELS, H.: 'Was ist "a priori" und was ist "a posteriori"?' *Philosophisches Jahrbuch der Görresgesellschaft*, 38 (1925), 201–210 and 321–332. For Bolzano, see 321–325.

(B38) GOTTHARDT, G.: *Bolzanos Lehre vom 'Satze an sich' in ihrer methodologischen Bedeutung.* Berlin, 1909, viii+120.

(B39) HEESCH, E.: *Grundzüge der Bolzanoschen Wissenschaftslehre*. Dissertation Münster (bei Scholz), 1933. Not seen by the Translator.

(B40) KREIBIG, J. C.: 'Ueber ein Paradoxon in der Logik Bolzanos.' *Vierteljahresschrift für wissenschaftliche Philosophie und Soziologie*, (1) 28=(2) 3 (1904), 375–391.

(B41) MAXSEIN, A.: *Die Entwicklung des Begriffs 'a priori' von Bolzano über Lotze zu Husserl und den von ihm beeinflussten Phänomenologen*. Dissertation Giessen, 1933. Not seen by the Translator.

(B42) PFEIFFER, F.: *Bolzanos Logik und das Transzendenzproblem*. Dissertation Zürich, 1922. Not seen by the Translator.

(B43) SCHOLZ, H.: *Geschichte der Logik*. Berlin 1931, vii+78. Geschichte der Philosophie in Längsschnitten, Heft 4.

(B44) ZIEHEN, TH.: *Lehrbuch der Logik auf positivistischer Grundlage mit Berücksichtigung der Geschichte der Logik*. Bonn, 1920, viii+866. For Bolzano, see 173–174.

ON BOLZANO THE MATHEMATICIAN

(B45) BERGMANN, H.: *Bolzanos Beiträge zur philosophischen Begründung der Mathematik* (B1, 155–212).

(B46) CANTOR, G.: 'Ueber unendliche lineare Punktmannichfaltigkeiten.' *Mathematische Annalen*, 21 (1883), 560–561 and 576. This is the passage on Bolzano; the whole contribution, the fifth of six, occupying pp. 545–591.

(B47) HANKEL, H.: *Grenze*. *Ersch und Grubers Allgemeine Enzyklopädie*, (1) 90 (1871), 185–211. For Bolzano, see 209–210.

(B48) JARNÍK, V.: 'O funkci Bolzanově.' *Časopis pro Pěstování matematiky a fysiky*, 51 (1922), 248–264 and 333–334.

(B49) JAŠEK, M.: 'Ueber den wissenschaftlichen Nachlass Bernhard Bolzanos.' *Jahresberichte der deutschen Mathematikervereinigung*, 31 (1922), 109–110 in italic pagination.

(B50) JAŠEK, M.: 'Funkce Bolzanova.' *Časopis pro Pěstování matematiky a fysiky*, 51 (1922), 69–76 and 325–326.

(B51) JAŠEK, M.: 'Aus dem handschriftlichen Nachlass Bernhard Bolzanos.' *Věstník Královské České Společnosti Nauk, Třída matematicko-přírodovědecká* (1923), Memoir I, pp. 1–32.

(B52) KORSELT, A.: 'Ueber die Grundlagen der Mathematik.' *Jahresberichte der deutschen Mathematikervereinigung*, 14 (1905), 365–389. For Bolzano, see pp. 365, 367, 373, 380, 387.

(B53) KOWALEWSKI, G.: 'Bolzanos Verfahren zur Herstellung einer nirgends differenzierbaren stetigen Funktion.' *Leipziger Berichte, mathematisch-physikalische Klasse*, 74 (1922), 91–95.

(B54) Kowalewski, G.: 'Ueber Bolzanos nichtdifferenzierbare stetige Funktion.' *Acta Mathematica*, 44 (1923), 315–319.

(B55) Rychlík, K.: 'Ueber eine Funktion aus Bolzanos schriftlichem Nachlass.' *Věstník Královské České Společnosti Nauk, Třída matematicko-přírodovědecká*, (1923), Memoir IV, pp. 1–20.

(B56) Schwarz, H. A.: 'Zur Integration der partiellen Differentialgleichung $\partial^2 u/\partial x^2 + \partial^2 u/\partial y^2 = 0$.' *Crelle*, 74 (1872), 218–253. For Bolzano, see footnote, pp. 220–221.

(B57) Stolz, O.: 'Bernhard Bolzanos Bedeutung in der Geschichte der Infinitesimalrechnung.' *Mathematische Annalen*, 18 (1881), 255–279.

ON BOLZANO THE PHILOSOPHER

For works up to 1909, see Bergmann (B1, 212–219). Only the following later works call for mention in this place.

(B58) Bergmann, H.: *Das philosophische Werk Bernard Bolzanos, mit Benutzung ungedruckter Quellen kritisch untersucht, nebst einem Anhange: Bolzanos Beiträge zur philosophischen Grundlegung der Mathematik*. Halle, 1909, xiv+230.

(B59) Fels, H.: 'Die Philosophie Bolzanos.' *Philosophisches Jahrbuch der Görresgesellschaft*, 40 (1927), 319–336 and 423–448.

(B60) Franzis, E.: *Bernhard Bolzano: der pädagogische Gehalt seiner Lehre, zugleich ein Beitrag zur Geistesgeschichte des ostmitteleuropäischen Raumes*. Münster in Westfalen, 1933, xx–249.

(B61) Gerhardt, J.: *Das Wahrheitsproblem in dem philosophischen Lebenswerke Bernhard Bolzanos*. Dissertation Münster, 1918; Trier, 1918, xli+122+33.

(B62) Staehler, W.: *Das Unsterblichkeitsproblem in der Philosophie Bernhard Bolzanos*. Fulda, 1928. Not seen by the Translator.

(B63) Waldschmitt, L.: *Bolzanos Begründung des Objektivismus in der theoretischen und praktischen Philosophie*. Würzburg, 1937, iv+117.

(B64) Winter, E.: *Religion und Offenbarung in der Religionsphilosophie Bernhard Bolzanos, dargestellt mit erstmaliger Heranziehung des handschriftlichen Nachlasses Bolzanos*. Breslau, 1932, 0+183. Breslauer Studien zur historischen Theologie, Band XX.

KEY TO THE QUOTATIONS IN THE PARADOXIEN

Bolzano's quotations were translated literally as they occurred, but not identified in individual footnotes. They are here collected in alphabetical order of authors, so as to present a picture of his source material. Such references as those to the passage in Kant on space and time as forms of intuition or to the passages in Leibniz on pre-established harmony are not named here, but remitted to the reference literature of philosophy.

FROM ARCHIMEDES, 287–212

In §37 and §40 to the Postulata in *De sphaero et cylindro*, ed. Heiberg, I, 8: 2–26.

FROM ARISTOTLE, 384–322

In §42 to the Corpus Aristotelicum, *Mechanica*, XXIV, 855a 28–856a 38; and *Problemata*, XVI, §5, 913b 37–914a 24.

FROM BOLZANO'S OTHER WORKS

In §7 to *Wissenschaftslehre*, §§82–87, pp. 393–414.
In §16 to *Wissenschaftslehre*, §87, pp. 409–414.
In §26 to *Wissenschaftslehre*, §46, p. 206.
In §27 to *Drei Dimensionen* generally; only 15 pp.
In §27 to *Zusammensetzung der Kräfte*, pp. 7–9 and 26–33.
In §55 to *Athanasia*, 1838 edition, pp. 133–134.
In §55 to *Wissenschaftslehre*, §79, pp. 361–378.
In §61 to *Athanasia*, pp. 104–105.
In §65 to *Athanasia*, pp. 132–133.

FROM J. R. BOSCOVICH, 1711–1787

In §44 to *Elementorum Universae Matheseos*, Tomus III, Rome, 1754, pp. 297–468.

In §70 to *De motu corporis attracti in centrum immobile viribus decrescentibus in ratione distantiarum reciprocata duplicata in spatiis non resistentibus*. Bononiense Scientiarum et Artium Instituto atque Academia Commentarii, 2/iii (1747), 262–288. See pp. 284–285.

KEY TO THE QUOTATIONS IN THE PARADOXIEN

FROM A. L. CAUCHY, 1789–1857

In §12 to *Cours d'Analyse de l'Ecole Royale Polytechnique.* Première Partie: Analyse algébrique, Paris 1821. Préliminaires. In the Oeuvres, (2) 3 (1897), 19.

In §37 to the same work, general reference.

FROM J. E. ERDMANN, 1805–1892

In §11 to *Grundriss der Logik und Metaphysik.* Zweite, verbesserte Auflage, Halle 1843, pp. 30–31.

FROM L. EULER, 1707–1783

In §36 to *Institutiones Calculi Differentialis*, Saint Petersburg and Berlin, 1755, §83. Opera (1) 10 (1913).

In §70 to *Mechanica sive motus scientia analytice exposita*, Saint Petersburg, 1736, Volume I, §655. Opera (2) 1 (1912), 218.

FROM J. K. FISCHER, 1760–1833

In §42 to *Grundriss der gesamten reinen höheren Mathematik.* Leipzig, 2 (1807), §51 Anmerkung.

FROM J. F. FRIES, 1773–1843

In §12 to *Die mathematische Naturphilosophie nach philosophischer Methode bearbeitet*, Heidelberg, 1822, §47, p. 255.

FROM G. GALILEI, 1564–1642

In §46 to *Discorsi e dimostrazioni matematiche intorno a due nuove scienze attenenti alla meccanica ed ai movimenti locali.* Leiden, 1638. In the Edizione Nazionale, 8 (1898), 73–77.

FROM J. A. GRUNERT, 1797–1872

In §12 to his article 'Unendlich' in G. S. Klügel's *Mathematisches Wörterbuch*, Teil 5 (T–Z, 1831) 507–537, in particular p. 513.

FROM J. F. HERBART, 1776–1841

In §40 to *Hauptpunkte der Metaphysik.* Göttingen, 1808, §7, pp. 46–58 and §9, p. 67. The two 'spaces' in question are the intelligible and the empirical.

FROM A. G. KAESTNER, 1719–1800

In §1 to *Anfangsgründe der Analysis des Unendlichen.* Göttingen, 1799, Vorrede, p. x.

In §46 to the same, pp. vii–ix.

KEY TO THE QUOTATIONS IN THE PARADOXIEN

FROM J. L. LAGRANGE, 1736–1813

In §37 Bolzano refers to the term 'une fonction dérivée.' It did not first appear in the *Théorie des fonctions*; the name occurs in *Sur une nouvelle espèce de calcul relatif à la différentiation et à l'intégration des quantités variables*, Nouveaux Mémoires de Berlin for 1772, 186–221; and Oeuvres (1) 3 (1869), 441–476. The notation occurs in the *Nouvelle méthode pour résoudre des équations littérales par le moyen des séries*, Mémoires de Berlin for 1768, 251–236; and Oeuvres (1) 3 (1869), 5–73. In particular, see in the former, §3; and in the latter, §16.

FROM G. W. LEIBNIZ, 1646–1716

In the Motto to *Réponse de M. Leibniz à l'extrait de la lettre de M. Foucher, Chanoine de Dijon, insérée dans le Journal du* 16 *mars* 1693. Gerhardt, Philosophische Schriften, 1 (1875) 416.

FROM M. R. S.

In §32 to 'Note sur quelques expressions algébriques peu connues.' Gergonne's *Annales de mathématiques pures et appliquées*, 20 (1830), 352–366. See in particular pp. 363–364.

FROM I. NEWTON, 1642–1727

In §40 to space as the 'sensorium Dei,' named in the later editions of the *Optics*, but not in 1704. It occurs near the end of the enlarged text, and in Horsley, Volume IV, p. 262.

FROM M. OHM, 1792–1872

In §34 to *Versuch eines vollkommen konsequenten Systems der Mathematik*, Berlin, 1822, in three volumes and eight parts. In Volume I, Edition 3, 1853, see §23.

FROM J. SCHULZ, 1739–1805

In §44 to *Versuch einer genauen Theorie des Unendlichen.* Erster Teil, vom Unendlichgrossen und der Messkunst derselben. Königsberg und Leipzig, 1788.

FROM B. SPINOZA, 1632–1677

In §12 to Spinoza's Letter to Ludwig Meyer, 20 April 1663. See J. van Vloten and J. P. N. Land's edition of the Opera quotquot reperta sunt, 's Gravenhaage, 1882–3, letter 'XII olim XXIX.'

If this letter by Spinoza is read more completely, it will be seen that Bolzano does not do him entire justice.

INDEX OF PERSONS AND PLACES

INDEX OF TOPICS

INDEX OF TOPICS